Fundamentals of Spread Spectrum Modulation

Fundamentals of Spread Spectrum Modulation
Rodger E. Ziemer

ISBN-978-3-031-00546-6 paperback
ISBN-978-3-031-00546-6 paperback

ISBN-978-3-031-01674-5 ebook
ISBN-978-3-031-01674-5 ebook

DOI: 10.1007/978-3-031-01674-5

A Publication in the Springer series
SYNTHESIS LECTURES ON COMMUNICATIONS #3

Lecture #3
Series Editor: William Tranter, Virginia Tech

Series ISSN: 1932-1244 print
Series ISSN: 1932-1708 electronic

First Edition
10 9 8 7 6 5 4 3 2 1

Printed in the United States of America

Fundamentals of Spread Spectrum Modulation

Rodger E. Ziemer
University of Colorado at Colorado Springs

SYNTHESIS LECTURES ON COMMUNICATIONS #3

ABSTRACT

This lecture covers the fundamentals of spread spectrum modulation, which can be defined as any modulation technique that requires a transmission bandwidth much greater than the modulating signal bandwidth, *independently* of the bandwidth of the modulating signal. After reviewing basic digital modulation techniques, the principal forms of spread spectrum modulation are described. One of the most important components of a spread spectrum system is the spreading code, and several types and their characteristics are described. The most essential operation required at the receiver in a spread spectrum system is the code synchronization, which is usually broken down into the operations of acquisition and tracking. Means for performing these operations are discussed next. Finally, the performance of spread spectrum systems is of fundamental interest and the effect of jamming is considered, both without and with the use of forward error correction coding. The presentation ends with consideration of spread spectrum systems in the presence of other users. For more complete treatments of spread spectrum, the reader is referred to [1, 2, 3].

KEYWORDS

Code acquisition, Code tracking, Direct sequence, Forward error correction coding, Frequency hop, Jamming, Multiple access noise, Receiver capture, Spread spectrum.

Contents

Fundamentals of Spread Spectrum Modulation

1 INTRODUCTION

A spread spectrum modulation scheme is any digital modulation technique that utilizes a transmission bandwidth much greater than the modulating signal bandwidth, *independently* of the bandwidth of the modulating signal.

There are several reasons why it might be desirable to employ a spread spectrum modulation scheme. Among these are to provide resistance to unintentional interference and multipath transmissions, to provide resistance to intentional interference (also known as jamming) [4], to provide a signal with sufficiently low spectral level so that it is masked by the background noise (i.e., to provide low probability of detection), and to provide a means for measuring range between transmitter and receiver.

Spread spectrum systems were historically applied to military applications and still are. Much of the literature on military applications of spread spectrum communications is classified. A notable application of spread spectrum to civilian uses was to cellular radio in the 1990s with the publication of interim standard IS-95 by the US Telecommunications Industry Association (TIA) [5]. Another more recent application of spread spectrum to civilian uses is to wireless local area networks (LANs), with standard IEEE 802.11 published under the auspices of the Institute of Electrical and Electronics Engineers (IEEE) [6]. The original legacy standard, released in July 1997, includes spread spectrum modem specifications for operation at data rates of 1 and 2 Mbps, and the 802.11b standard, released in Oct. 1999, has a maximum raw data rate of 11 Mbps with both operating in the 2.4 GHz band. Specifications 802.11a and 802.11g, released in Oct. 1999 and June 2003, respectively, use another modulation scheme known as orthogonal frequency division multiplexing, with the former operating in the 5 GHz band and the latter operating in the 2.4 GHz band.

The schematic diagram shown in Fig. 1 may be used to explain several features of a spread spectrum modulation system. The type of spread spectrum system shown in Fig. 1 is known as *binary direct sequence* (DS) spread spectrum modulation, because a data bit 1 (of duration T_b) is sent as the spreading code, $c_1(t)$, noninverted and a data bit 0 (of duration T_b) is sent as the spreading code inverted or negated. (A spreading code is a repeating sequence of $N \pm 1$-s each T_c seconds in duration, called *chips*, produced by a feedback digital circuit.) Two practices regarding

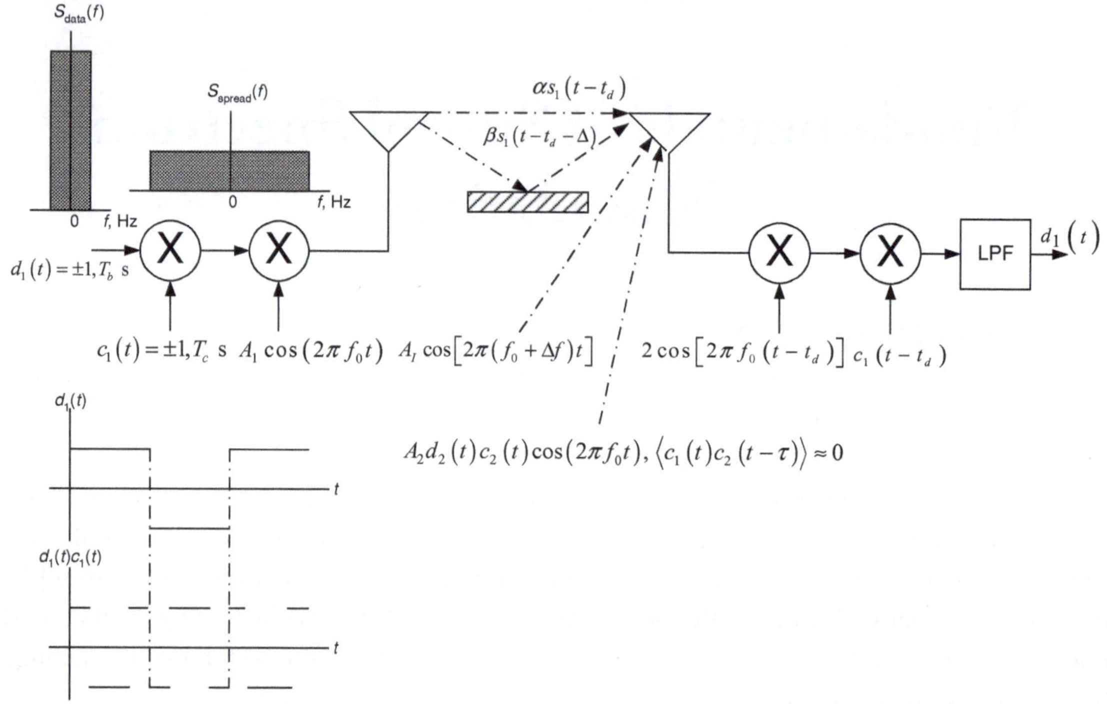

FIGURE 1: A basic spread spectrum communication system showing several possible channel impairments.

the spreading code in a DS system are commonly used: (1) all N chips of the code are contained in 1-bit interval ($NT_c = T_b$) (called a short code system) and (2) the spreading code is several data bits long (called a long code system). We assume the former in this discussion for simplicity. Because of the multiplication of each data bit by the spreading code, whose chip durations are T_b/N, the spectrum of the signal, i.e., of $d_1(t)\,c_1(t)$, is spread beyond the bandwidth of $d_1(t)$ by a factor of N. The final operation at the transmitter is to multiply the spread data signal by the carrier to produce the transmitted spread spectrum signal $s_1(t) = A_1 d_1(t)\,c_1(t)\cos(2\pi f_0 t)$. This signal propagates to the antenna of the receiver and arrives as $\alpha s_1(t - t_d)$, being both attenuated by a factor α and delayed by t_d s. It is assumed that the receiver can produce replicas of both the carrier, $2\cos[2\pi f_0(t - t_d)]$ (the factor 2 is for convenience), and the code, $c_1(t - t_d)$. Producing either of these is easy—the first simply takes a relatively stable oscillator and the latter takes the same feedback digital structure as used at the transmitter. The trick is to get both into synchronism with the incoming signal—a process called synchronization and tracking for which there are solutions. Assuming that this has been accomplished successfully, the steps in the receiver are to multiply by the locally generated carrier and code and then lowpass filter the result. The product $2\alpha A_1 d_1(t - t_d)\,c_1^2(t - t_d)\cos^2[2\pi f_0(t - t_d)]$ simplifies to $\alpha A_1 d_1(t-t_d)\{1+\cos[4\pi f_0(t-t_d)]\}$ because $c_1^2(t-t_d)=1,\ 0\le t\le T_b$, and $2\cos^2 x = 1+\cos(2x)$. Thus, the lowpass filter output is $\alpha A_c d_1(t-t_d)$.

Several other signals are shown entering the antenna of the receiver in Fig. 1. First, there is the signal $\beta s_1(t-t_d-\Delta)$, which represents the transmitted signal reflected from another object and is commonly called a *multipath* signal component. Having come from an indirect path to the receiver antenna, it has a delay, Δ, in addition to the delay of the direct-path signal. When multiplied by the locally generated carrier and code references in the receiver, the result is $2\beta A_1 d_1(t-t_d)c_1(t-t_d)c_1(t-t_d-\Delta)\cos[2\pi f_0(t-t_d)]\cos[2\pi f_0(t-t_d-\Delta)]$. Now the spreading codes are chosen so that the average of the product $c_1(t-t_d)c_1(t-t_d-\Delta)$ is small for $|\Delta|>T_c$, so this term is discriminated against by the receiver. Another signal component present at the receiver input is shown as $s_2(t)=A_2 d_2(t)c_2(t)\cos(2\pi f_0 t)$ and represents a signal transmitted by another user. In a spread spectrum system, the codes are chosen from a code family with the property that $\langle c_1(t)c_2(t-\tau)\rangle\approx 0$ where the angular brackets, $\langle\,\rangle$, represent the time averaging performed by the lowpass filter. Thus, signals broadcast by other transmitters will be discriminated against if the spreading codes are chosen properly. Finally, there is the signal $A_I\cos[2\pi(f_0+\Delta f)t]$, which represents a narrowband interfering signal, either intentional or unintentional. When this signal enters the receiver, it will be multiplied by the locally generated code, $c_1(t-t_d)$, and the resulting signal will be *spread* in bandwidth with a spectral level that is correspondingly reduced. Thus, much less power from this signal will be passed by the lowpass filter than if it had not been spread by the local code. In other words, the receiver will discriminate against narrowband interference present at its input. The ratio $G_p=T_b/T_c$ is also the ratio of spread bandwidth to data bandwidth and is called the *spreading factor* or the *processing gain*. The processing gain is a measure of the amount of discrimination provided against interfering signals.

2 REVIEW OF BASIC DIGITAL MODULATION TECHNIQUES

Before getting into the details of spread spectrum modulation schemes, it will be useful for future reference to review basic digital modulation techniques. The block diagram of Fig. 2 shows the basic idea. The receiver block is labeled "maximum likelihood" to denote a receiver which observes the received signal plus noise over a T_s-second interval and chooses the signal that is most likely to have resulted in the observed data. We have a source, which for simplicity will be assumed to have a binary alphabet (say $\{0,1\}$) that is composed of characters, or bits, each T_m seconds. This bit stream is to be associated in a unique fashion with a sequence of waveforms, each of duration T_s, from the set $\{s_0(t), s_1(t), \ldots, s_{M-1}(t)\}$. Clearly, if $M=2$, a useful association is $0\rightarrow s_0(t);\ 1\rightarrow s_1(t)$ while, if $M=4$, a useful association might be $00\rightarrow s_0(t),\ 10\rightarrow s_1(t),\ 11\rightarrow s_2(t),\ 01\rightarrow s_3(t)$ (other associations are clearly possible).

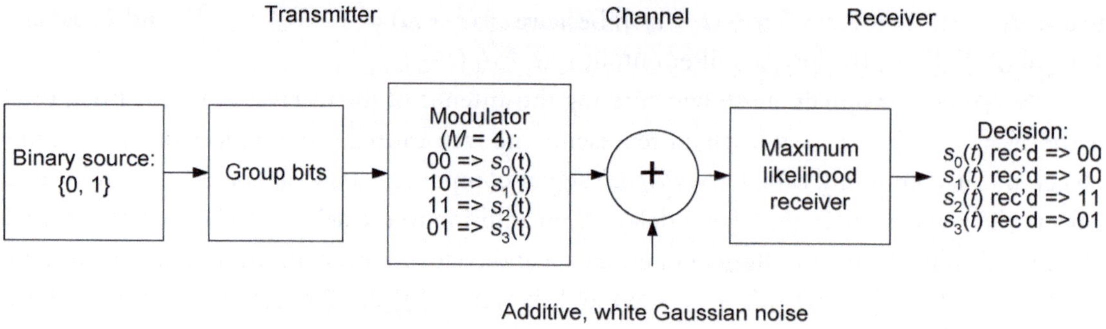

FIGURE 2: Block diagram of an M-ary digital transmission system ($M = 4$ used for illustration).

In both examples, if no gaps are to be present in the character or signal sequences, it must be true that $(\log_2 M)\, T_m = T_s$. In terms of rate, we have

$$R_m = (\log_2 M)\, R_s\,, \tag{2.1}$$

where $R_m = 1/T_m$ characters (bits) per second and $R_s = 1/T_s$ symbols per second.

Things are a bit more complicated if the source alphabet is not binary, but such cases will not be needed in this discussion. We call a modulation scheme selecting one of M possible signals to transmit each T_s-seconds M-ary, with the case of $M = 2$ referred to simply as a binary scheme. Table 1 gives a few examples of M-ary signaling schemes.

A digital modulation scheme is coherent or noncoherent depending on whether the received signal is demodulated by means of a local carrier in phase coherence with the received signal or not. For a coherent receiver, the general form for an M-ary communication receiver is a parallel matched filter, or correlator, bank (one for each possible transmitted signal) followed by a decision box. By expressing the possible transmitted signals as linear combinations of a set of K functions orthogonal over $[0,\ T_s]$ (always possible using the Gram–Schmidt procedure), this number, M, of correlators can be reduced to $K \leq M$. For a noncoherent receiver, a method of detection not dependent on signal phase must be used. For the M-ary FSK case, this involves a bank of $2M$ correlators (or matched filters), one for a cosine and one for a sine carrier reference for each possible transmitted signal, a squarer at each output, a bank of M summers, and a decision box.

The two primary performance measures of interest for a digital modulation scheme are its bandwidth efficiency and its communication efficiency. The former is characterized by the ratio of bit rate to some measure of bandwidth (often the null-to-null bandwidth of the main lobe of its signal spectrum for simplicity). Since both rate and bandwidth have the dimensions of inverse seconds, this ratio is, strictly speaking, dimensionless but the dimensions are usually referred to as bits per second per hertz (bps/Hz). The communications efficiency is measured by

TABLE 1: Signal Sets for Some Digital Modulations Schemes

NAME OF MODULATION SCHEME	SIGNAL SET: $0 \leq t \leq T_s$
Binary phase-shift keying (BPSK)	$1 \Rightarrow A_c \cos(2\pi f_0 t)$; $f_0 = n/T_s$, $T_s = T_b$, n integer $0 \Rightarrow -A_c \cos(2\pi f_0 t)$
Binary differential phase-shift keying (DPSK)	Binary bit stream differentially encoded (DE); DE bits BPSK modulate the carrier. DE: data-bit 1 encoded as no change from reference bit; data-bit 0 encoded as a change from reference bit; current encoded bit is reference for next encoded bit.
Binary frequency-shift keying (BFSK)	$A_c \cos(2\pi f_0 t)$, $A_c \cos[2\pi(f_0 + \Delta f)t]$; $f_0 = n/T_b$, $\Delta f = m/T_b$, m, n integers, $m \neq n$
M-ary phase-shift keying (MPSK) $M = 4$ called quadriphase-shift keying (QPSK)	$A_c \cos(2\pi f_0 t + 2(i-1)\pi/M)$, $i = 1, 2, \ldots, M$ $f_0 = n/T_s$, n integer
M-ary frequency-shift keying (MFSK)	$A_c \cos[2\pi(f_0 + (i-1)\Delta f)t]$; $f_0 = n/T_s$, $\Delta f = m/2T_s$ ($m \geq 1$ for orthogonal signals)

a communication system's performance in terms of bit error probability versus signal-to-noise ratio, usually specified as E_b/N_0, where E_b is the bit energy for the signal ($E_b = E_s/\log_2 M$ for an M-ary system, where E_s is the symbol energy) and N_0 is the one-sided power spectral density of the white, Gaussian background noise at the receiver input. Table 2 summarizes the bandwidth and communications efficiencies in additive white Gaussian noise (AWGN) for various digital modulation schemes.

In the preceding discussion, it was presumed that the channel imposes a fixed attenuation and the only signal impairment was the AWGN at the receiver input (modeled as entering the system at this point because that is where the signal is weakest). Another common channel model is the one with time varying attenuation, perhaps due to obstructions or reflections, of the signal. If these attenuation variations are slow enough, they can be viewed as fixed throughout a

TABLE 2: Bandwidth and Communication Efficiencies of Some Digital Modulation Schemes

NAME OF MODULATION SCHEME	BANDWIDTH EFFICIENCY (BPS/HZ)	BIT ERROR PROBABILITY, P_b	E_b/N_0 REQUIRED FOR $P_b = 10^{-6}$
Binary phase-shift shift keying (BPSK)	0.5	$Q\left(\sqrt{2E_b/N_0}\right)^a$	$E_b/N_0 = 10.53$ dB
Binary frequency-shift keying (BFSK)	0.4 coherent, 0.25 noncoherent	$Q\left(\sqrt{E_b/N_0}\right)$ coherent, $0.5\exp\left[-E_b/(2N_0)\right]$ noncoherent	13.54 dB coherent 14.2 dB noncoherent
Binary differential phase-shift keying (DPSK)	0.5	$0.5\exp\left(-E_b/N_0\right)$ (DE bit stream, see Table 1; 0 sent as π-'rad phase shift; 1 sent as 0 rad phase shift)	11.2 dB
M-ary DPSK	$0.5\log_2 M$	$\approx \frac{2}{m}\sqrt{\frac{1+\cos(\pi/M)}{2\cos(\pi/M)}} \times Q\left(2m\left[1-\cos\left(\frac{\pi}{M}\right)\frac{E_b}{N_0}\right]\right)$ $m = \log_2 M, M > 2$	11.2 dB, $M = 2$ 12.9 dB, $M = 4$ 16.8 dB, $M = 8$ 21.4 dB, $M = 16$ 26.3 dB, $M = 32$
M-ary phase-shift keying (MPSK)	$0.5\log_2 M$	$\approx Q\left(\sqrt{2E_b/N_0}\right)$; $M = 2, 4$ $< 2Q\left(\sqrt{2m\left(\frac{E_b}{N_0}\right)}\sin\frac{\pi}{M}\right)$ $m = \log_2 M$ (bound tight for $M > 4$)	10.5 dB, $M = 2, 4$ 14 dB, $M = 8$ 18.5 dB, $M = 16$ 23.4 dB, $M = 32$ 28.5 dB, $M = 64$
M-ary frequency-shift keying (MFSK)	$\frac{2\log_2 M}{M+3}$ coherent $\frac{\log_2 M}{2M}$ noncoherent	$\frac{M}{2}Q\left(\sqrt{\log_2 M\left(\frac{E_b}{N_0}\right)}\right)$ coherent, $\frac{M}{2(M-1)}\sum_{k=1}^{M-1}\frac{(-1)^{k+1}}{k+1}\binom{M-1}{k} \times \exp\left[\frac{-k\log_2 M}{k+1}\frac{E_b}{N_0}\right]$ noncoherent	13.5 dB, $M = 2$ 10.8 dB, $M = 4$ 9.3 dB, $M = 8$ coherent 14.2 dB, $M = 2$ 11.4 dB, $M = 4$ 9.9 dB, $M = 8$ noncoherent

[a] $Q(x) = \int_x^\infty \frac{\exp(-u^2/2)}{\sqrt{2\pi}}du = \int_0^{\pi/2}\exp\left(-\frac{u^2}{2\sin^2\phi}\right)\frac{d\phi}{\pi} \simeq \frac{\exp(-x^2/2)}{x\sqrt{2\pi}}$, $x > 4$.

given signaling interval. Perhaps the most frequently used model is the slow flat Rayleigh fading model, wherein a given transmitted signal is attenuated by a fixed (for that symbol interval) level modeled as a Rayleigh-distributed random variable and the next transmitted signal is likewise attenuated by a new, independent Rayleigh random variable, etc. For sufficiently slow fading, this model can be a fairly accurate representation of the true state of affairs, and it is simple to analyze a digital transmission system experiencing such a channel. The analysis proceeds by using the error probability expressions of Table 2 and averaging over the signal-to-noise ratio, E_b/N_0, not with respect to a Rayleigh probability distribution, but with respect to an exponential probability distribution because $E_b/N_0 = A_c^2 T_b/2N_0$, where A_c is the signal amplitude which is modeled as a Rayleigh random variable. Thus, E_b, being proportional to the signal amplitude squared, is exponentially distributed. This results in a particularly simple integral to evaluate in the case of binary DPSK or NFSK. For the latter case,

$$\overline{P}_{b,\mathrm{NFSK}} = \int_0^\infty \frac{1}{2} \exp\left(-z/2\right) \frac{1}{\overline{Z}} \exp\left(-z/\overline{Z}\right) \mathrm{d}z = \frac{1}{2+\overline{Z}}, \tag{2.2}$$

where $\overline{Z}$ is the average received E_b/N_0. For DPSK, the integration is similar. For BPSK the integral is challenging but possible to perform. The results for these two cases are

$$\overline{P}_{b,\mathrm{DPSK}} = \frac{1}{2\left(1+\overline{Z}\right)}; \qquad \overline{P}_{b,\mathrm{BPSK}} = \frac{1}{2}\left(1 - \sqrt{\frac{\overline{Z}}{1+\overline{Z}}}\right). \tag{2.3}$$

The sobering fact about the effects of Rayleigh fading is the penalty imposed on communications efficiency. The difference between signal-to-noise ratios for fading and nonfading cases for a given modulation scheme is called the *fading margin* for that scheme. For a bit error probability of 10^{-3}, the fading margins for binary NFSK, DPSK, and BPSK are 16.04 dB, 19.05 dB, and 20.19 dB, respectively. For MPSK with $M = 8$ and 16, the fading margins are 15 dB and 14.6 dB, respectively. The question of what do about the penalty imposed by fading has a partial answer in the use of diversity, that is, providing several alternative paths to pass the signal through, not all of which will fade simultaneously, hopefully.

3 TYPES OF SPREAD SPECTRUM MODULATION

The two most common types of spread spectrum modulation are direct-sequence and frequency-hop spread spectrum (FHSS). A binary direct-sequence spread spectrum (DSSS) scheme was used in the illustrations of Fig. 1.

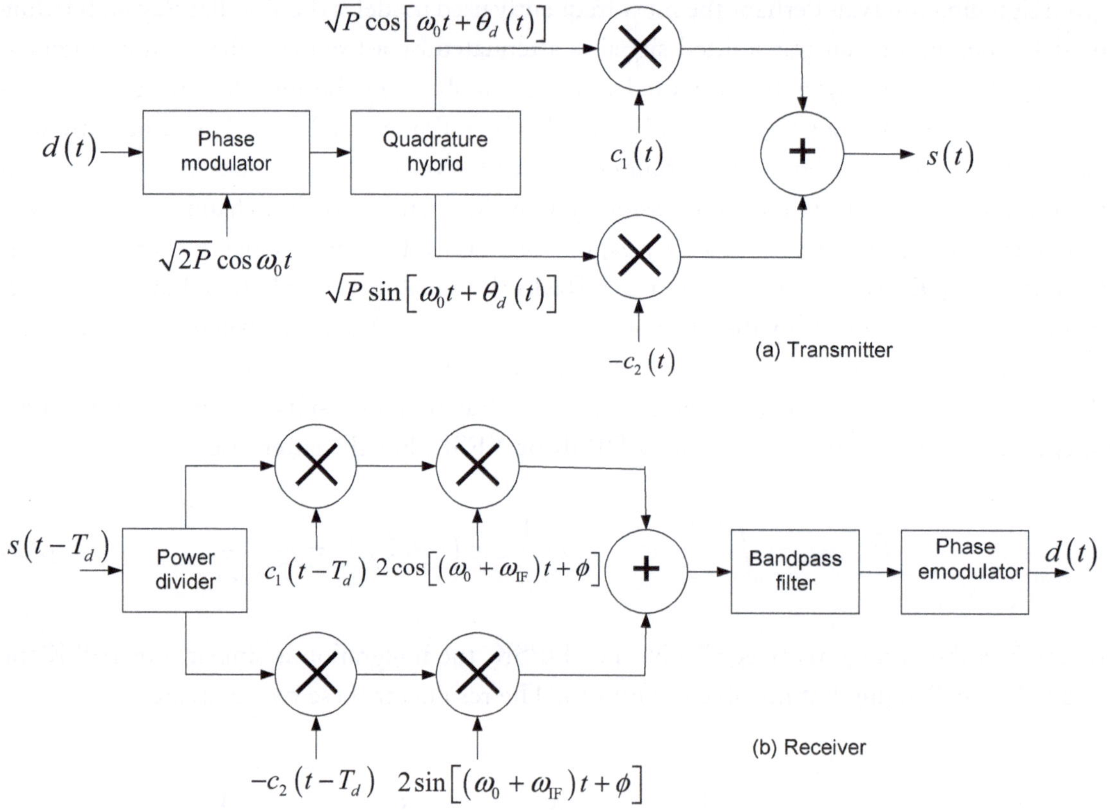

FIGURE 3: Block diagrams of the transmitter (a) and receiver (b) for QPSK spreading with arbitrary phase modulation [1].

3.1 QPSK Spreading With Data Phase Modulation

Modulation types other than BPSK may be used in DSSS communication systems, both for the data and for the spreading. For example, Fig. 3 shows a transmitter/receiver structure for QPSK spreading with arbitrary data phase modulation.

3.2 Frequency-Hop Spread Spectrum

As its name implies, FHSS involves hopping the data-modulated carrier pseudo-randomly in frequency. A combination of direct sequence and frequency hop modulation is often referred to as hybrid spread spectrum modulation. Another type of spread spectrum modulation, called time-hopped or pulse-position-hopped [3], involves time hopping the transmitted data pulses pseudo-randomly in time with respect to a fixed reference position for each signaling interval. While not prevalently implemented in the past, this type of spread spectrum is more popular recently because of the current intense exploration of ultra-wideband modulation techniques.

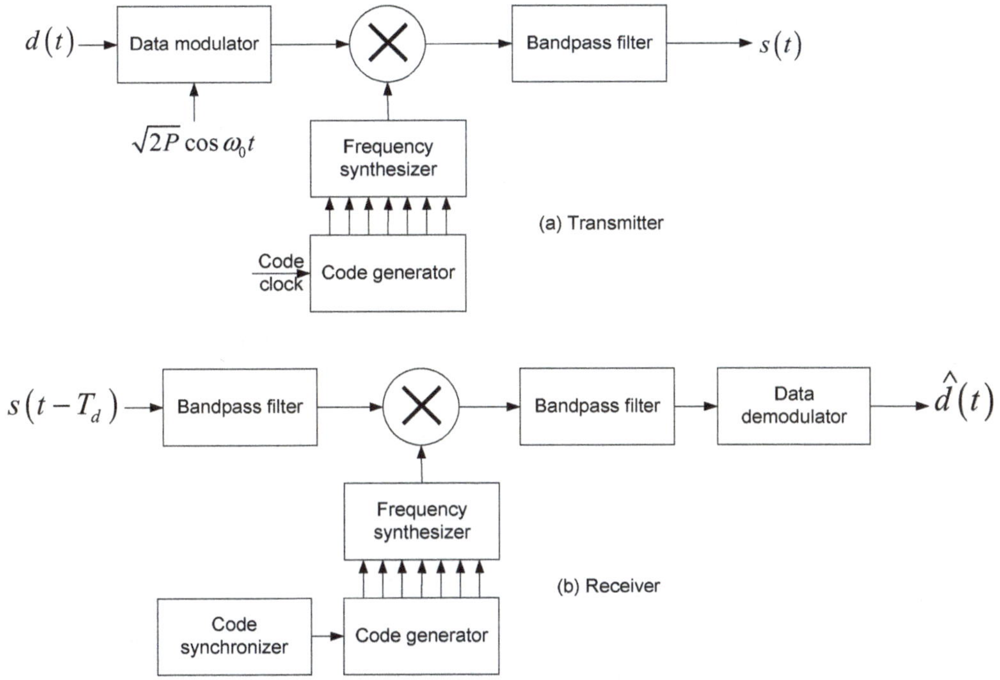

FIGURE 4: Block diagram of a FHSS transmitter (a) and receiver (b) [1].

The focus of attention in this section is on FHSS modulation since the idea of DSSS was explained in relation to Fig. 1. A schematic block diagram of a FHSS communication system is shown in Fig. 4. Often, a noncoherent data modulation scheme, such as noncoherent FSK or DPSK, is used since it is more difficult to build frequency synthesizers that maintain phase coherence from hop to hop than those that do not. A pseudo-random code generator is used as a driver for a frequency synthesizer at the transmitter to pseudo-randomly hop the carrier frequency of the data modulator output. In keeping with the basic idea of spread spectrum, the hopping frequency range is quite broad compared with the modulated data bandwidth. The time interval of a frequency hop is called the hop period, T_h. Two situations can prevail: the hop period can be *long* with respect to a data bit period; the hop period can be *short* with respect to a data bit period. The former case is referred to as *slow frequency hop*, and the latter case is referred to as *fast frequency hop*. Perhaps the most common situation in practice is slow frequency hop. Fast frequency hop has some advantages over slow frequency hop but is more difficult to implement.

At the receiver, a pseudo-random code generator identical to the one used at the transmitter is implemented and used to drive a frequency synthesizer like the one used at the transmitter. Assuming that the pseudo-random number sequence output by the number generator can be synchronized with the one at the transmitter (accounting for channel delay), the frequency hopping sequence will track that of the transmitted hopping sequence and the received frequency-hopped spread spectrum signal will be de-hopped whereupon an appropriate data demodulator can be used to recover the data sequence. In the early days of spread spectrum, FHSS was used to realize wider spread bandwidths than possible with DSSS systems.

If the features of FHSS and DSSS are combined, the result is referred to as *hybrid* spread spectrum. Usually, the additional implementation complexity does not warrant the hybrid approach, so the actual use of such systems is seen very little. One advantage of the hybrid approach is to force a potential hostile interceptor to use a more complex interception strategy [4].

Example 1. A binary data source emits binary data at a rate of $R_b = 10$ kbps. Each bit is sent in a DSSS communication system either as a 127-chip code as is (data bit 1) or inverted (data bit 0).

(a) What is the bandwidth of the DSSS transmitted signal?

(b) Compare this with a FHSS system that uses binary NFSK modulation. How many frequency hop slots are required to provide roughly the same transmission bandwidth as the DSSS system?

Solution:

(a) From Table 2, the bandwidth efficiency of BPSK is 0.5 which means that the transmission bandwidth of the unspread signal is $10/0.5 = 20$ kHz. The spread signal bandwidth is 127 times of this or 2.54 MHz.

(b) From Table 2, the bandwidth efficiency of binary NFSK is 0.4 which gives a transmission bandwidth for the unspread signal of $10/0.4 = 25$ kHz. The number of frequency hops needed to provide the same spread bandwidth as the DSSS system is $2,540,000/25,000 = 101.6$ which is rounded to 102. The spread bandwidth of the FHSS system is therefore 2.55 MHz, which is close to that of the DSSS system.

3.3 Summary

The previous two sections have laid the ground work for the consideration of spread spectrum communication systems with discussions of the basic idea of a direct sequence spread spectrum system and some of its features, a review of basic digital modulation techniques and, in addition to the DSSS system example, descriptions of two generic spread spectrum modulation techniques—QPSK spreading and frequency-hop spread spectrum.

4 SPREADING CODES

An important component of any spread spectrum system is the pseudo-random spreading code. Many options exist for the generation of such codes, only a few of which will be described here. In particular, m-sequences will first be described in terms of their generation and properties. Then, Gold codes will be discussed in terms of their generation and cross-correlation properties. Next, the small set of Kasami sequences will be introduced, followed by an introduction to quaternary sequences. Finally, the construction of Walsh functions will be illustrated.

4.1 Generation and Properties of m-Sequences

Maximal-length sequences, or m-sequences, are simple to generate with linear feedback shift-register circuits and have many nice properties. But, they are relatively easy to intercept and regenerate by an unintended receiver. While the theory of m-sequences cannot be discussed in detail here, two circuits for their generation will be given and some of their properties listed.

Figure 5 illustrates two feedback shift-register configurations for the generation of m-sequences. Each box represents a storage location for a binary digit, labeled with a D for delay

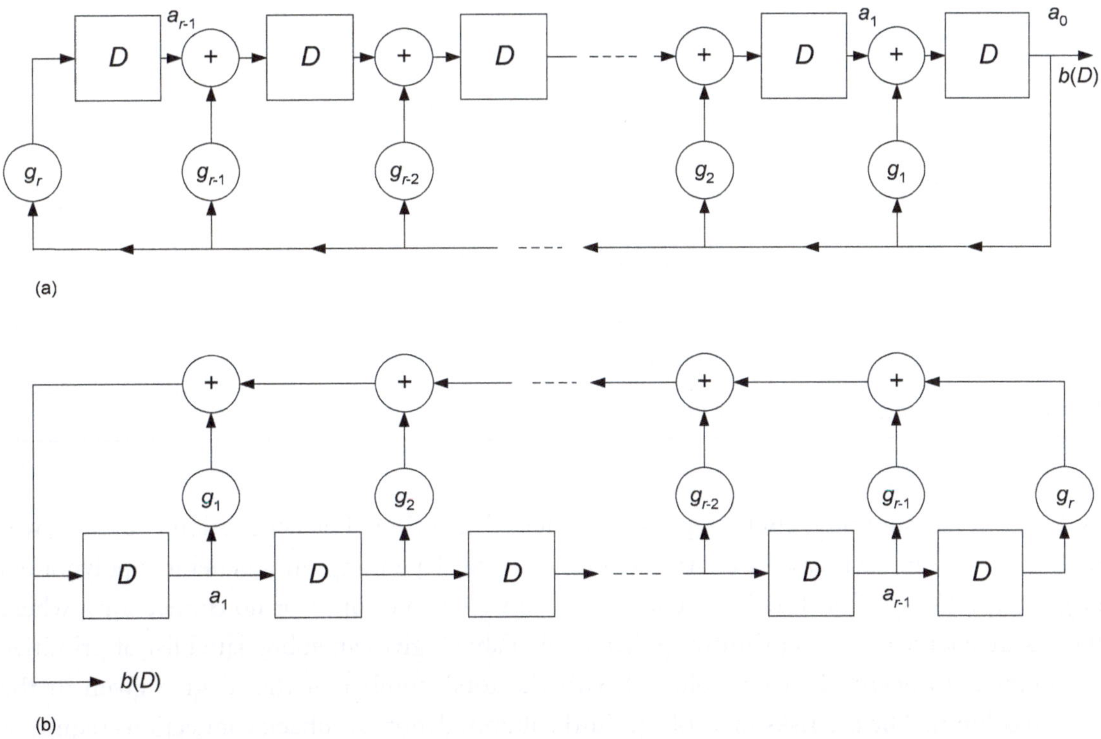

FIGURE 5: Two configurations of m-sequence generators: (a) high-speed linear feedback shift-register generator; (b) low-speed linear feedback shift-register generator [1].

TABLE 3: Some Generator Polynomial Coefficients in Octal Format for m-Sequences; $m = 2^r - 1$.

DEGREE, r	NO. OF PRIMITIVE POLYNOMIALS	OCTAL REPRESENTATION OF THE GENERATOR POLYNOMIAL (g_0 ON THE RIGHT; g_r ON THE LEFT)
2	1	[7]*
3	2	[13]*
4	2	[23]*
5	6	[45]*, [75], [67]
6	6	[103]*, [147], [155]
7	18	[211]*, [217], [235], [367], [277], [325], [203]*, [313], [345]
8	16	[435], [551], [747], [453], [545], [537], [703], [543]
9	48	[1021]*, [2231], [1461], [1423], [1055], [1167], [1541] [1333], [1605]
10	60	[2011]*, [2415], [3771], [2157], [3515], [2773], [2033], [2443], [2461]
11	176	[4005]*, [4445], [4215], [4055], [6015], [7413], [4143], [4563], [4053]

* Feedback connections from one intermediate delay.

by T_c s, and the summing circles represent modulo-2 addition. The connection circles, shown with a label g_i in each case, are either closed or open depending on a generator polynomial $g_r g_{r-1} \ldots g_0$ (1 signifies closed or a connection and 0 signifies open or no connection), where the g_is are coefficients of a primitive polynomial. Table 3 gives an abbreviated list of primitive polynomials to degree 11 (first column) with the total number of that degree given in the second column. The asterisks in Table 3, third column, denote feedback connections requiring only one adder. There are extensive tables of primitive polynomial coefficients to much higher degree [1].

In Table 3, the primitive polynomial coefficients are given in octal format. For example, taking the first entry of the degree 10 listing, we have

$$[2011]_8 \Leftrightarrow [010000001001]_2 \Leftrightarrow D^{10} + D^3 + 1. \tag{4.1}$$

All we want are the binary coefficients so that we know if a given connection is present or not in the shift-register circuits of Fig. 5. The particular 1s and 0s occupying the shift register stages after a clock pulse occurs are called *states*.

Example 2. An m-sequence of degree 3 is desired. Give the generator polynomial, the number of shift register stages, and the connections for the configurations of Fig. 5(a) and 5(b).

Solution: From Table 3, the generator octal and binary representations and generator polynomial are

$$[1\ 3]_8 = [0\ 0\ 1\ 0\ 1\ 1]_2 \Leftrightarrow D^3 + D + 1 = \sum_{i=0}^{r-1} g_i D^i.$$

The two generic forms of the sequence generators shown in Fig. 5 are specialized to this example and are shown in Fig. 6. Both generic forms have three delays in this example. Note that an initial load of 001 is assumed for the shift register of (a); subsequent states may then be found.

The following properties apply to m-sequences:

1. An m-sequence contains one more 1 than 0.
2. The modulo-2 sum of an m-sequence and any phase shift of the same m-sequence is another phase of the same m-sequence (a phase of the sequence is a cyclic shift).
3. If a window of width r is slid along an m-sequence for N shifts, each r-tuple except the all-zeros r-tuple will appear exactly once.

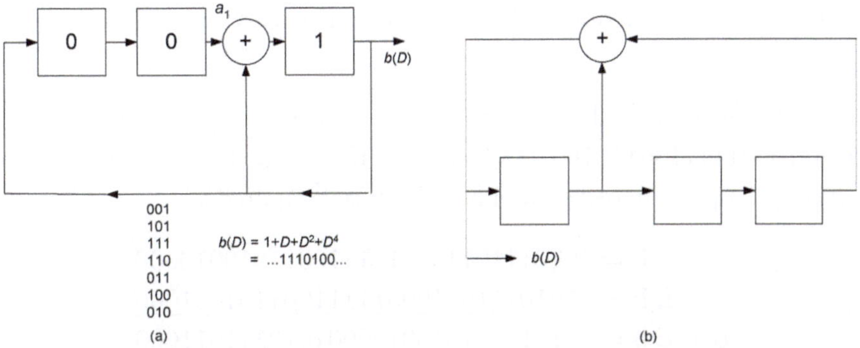

FIGURE 6: The two m-sequence shift-register configurations for Example 2.

4. Define a run as a subsequence of identical symbols within the m-sequence. Then, for any m-sequence, there are
   - One run of ones of length r.
   - One run of zeros of length $r-1$.
   - One run of ones and one run of zeros of length $r-2$.
   - Two runs of ones and two runs of zeros of length $r-3$.
   - Four runs of ones and four runs of zeros of length $r-4$.
   - ...
   - 2^{r-3} runs of ones and 2^{r-3} runs of zeros of length 1.
5. The autocorrelation function of a repeated m-sequence is periodic with period $T_0 = NT_c$ and is given by (0s replaced by -1 values)

$$R_c(\tau) = \frac{1}{T_0} \int_{T_0} x(t)x(t+\tau)\,\mathrm{d}t = \begin{cases} \left(1 - \frac{|\tau|}{T_c}\right)\left(1 + \frac{1}{N}\right) - \frac{1}{N}, & |\tau| \leq T_c \\ -\frac{1}{N}, & T_c < |\tau| \leq \frac{N-1}{2} T_c, \end{cases} \tag{4.2}$$

   where the integration is over any period, $T_0 = NT_c$.
6. The Fourier transform of the autocorrelation function of an m-sequence, which gives the power spectral density, is given by

$$S_c(f) = \sum_{m=-\infty}^{\infty} P_m \delta(f - mf_0), \; f_0 = 1/NT_c, \tag{4.3}$$

   where

$$P_m = \begin{cases} \left[(N+1)/N^2\right] \mathrm{sinc}^2(m/N), \, m \neq 0, \, \mathrm{sinc}(x) = (\sin \pi x)/(\pi x) \\ 1/N^2, \, m = 0. \end{cases}$$

The autocorrelation function and power spectral density of a 15-chip m-sequence are shown in Fig. 7.

Example 3. Consider the $2^5 - 1 = 31$-chip m-sequence:
$\mathbf{b} = 1010111011000111110011010010000$. We see that it has 16 1s and 15 0s (property 1).

The chip-by-chip modulo-2 sum of $\mathbf{b}$ and $D\mathbf{b}$ is computed as

$$\begin{aligned} \mathbf{b} &= 1010111011000111110011010010000 \\ D\mathbf{b} &= \underline{0101011101100011111001101001000} \\ \mathbf{b} + D\mathbf{b} &= 1111100110100100001010111011000 \end{aligned}$$

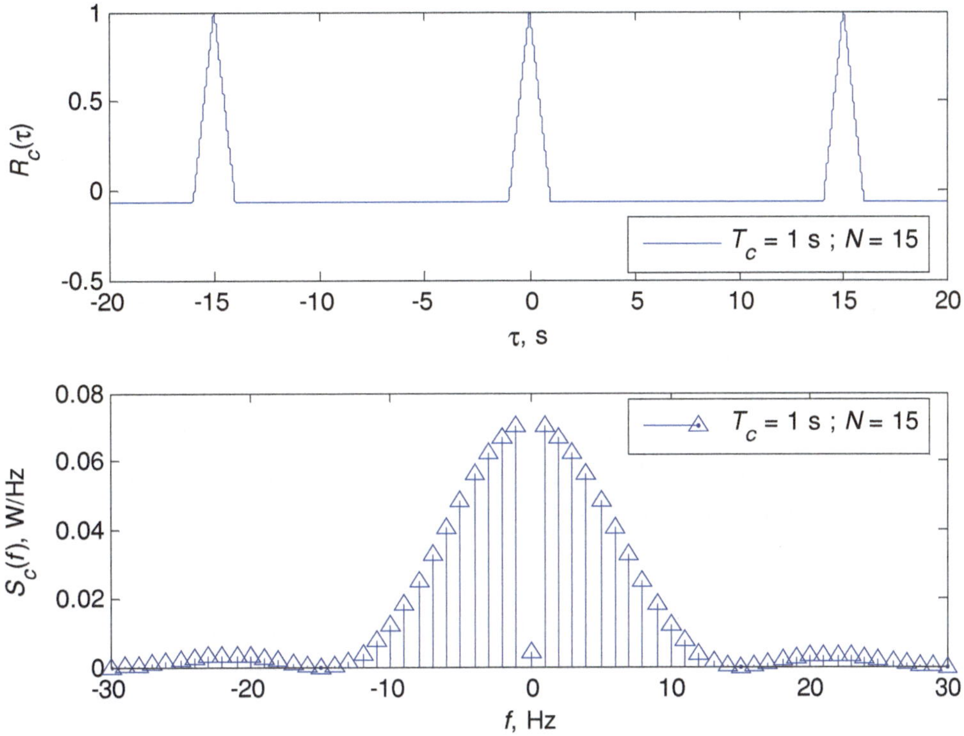

FIGURE 7: Autocorrelation function (top) and power spectral density (bottom) of an m-sequence.

which is seen to be a 13-chip shift of $\mathbf{b}$ (property 2).

Taking a window of width $r = 5$ and sliding it along $\mathbf{b}$ (periodically extended) gives the 5-tuples 10101, 01011, 10111, . . . , 10000, 00001, 00010, 00101, 01010 (31 total). An extended listing shows that all possible 5-tuples are present, with the exception of 00000 (property 3). Close examination of the sequence $\mathbf{b}$ shows that there are the following runs:

- One run of 1s of length $r = 5$;
- One run of 0s of length $r - 1 = 4$;
- One run of 1s and one run of 0s of length $r - 2 = 3$;
- Two runs of 1s and two runs of 0s of length $r - 3 = 2$;
- Four runs of 1s and four runs of 0s of length $r - 4 = 1$ (property 4).

Property 5 follows by considering the autocorrelation function at delays equal to integer multiples of a chip and noting that the autocorrelation values between these delays must be a linear function of the delay. For $\tau = 0$, we get $R_c(0) = \frac{1}{T_0} \int_{T_0} x^2(t)\,dt = \frac{31T_c \times 1}{31T_c} = 1$. For a delay of

T_c, there is one more $1 \times (-1)$ value so the result is $R_c(T_c) = \frac{-T_c}{31T_c} = -\frac{1}{31}$, which holds for delays of $\pm 2T_c, \pm 3T_c, \ldots, \pm 15T_c$. For delays between these values, the autocorrelation function must, of necessity, be a linear function of τ (the integrand involves constants). Because the sequence is periodically extended, the autocorrelation function is also periodic of period $31T_c$.

Note that the correlation function given by (4.2) is obtained only if integration is over a full period. In spread spectrum systems, the correlation function of m-sequences when integrated over less than a period is important, especially for long codes. Although beyond the scope of this presentation, partial-period correlation values for m-sequences can be shown to be highly variable and not the nice result given by (4.2) [1].

The power spectrum of $b(t, \varepsilon) = c(t)c(t + \varepsilon)$ is an important consideration for synchronization. This is a complex problem [1]. Example results are shown in Fig. 8 where it is seen that significant power is at DC if $\varepsilon \leq T_c/2$; this is important because it is this component on which the tracking loop of a code synchronizer locks.

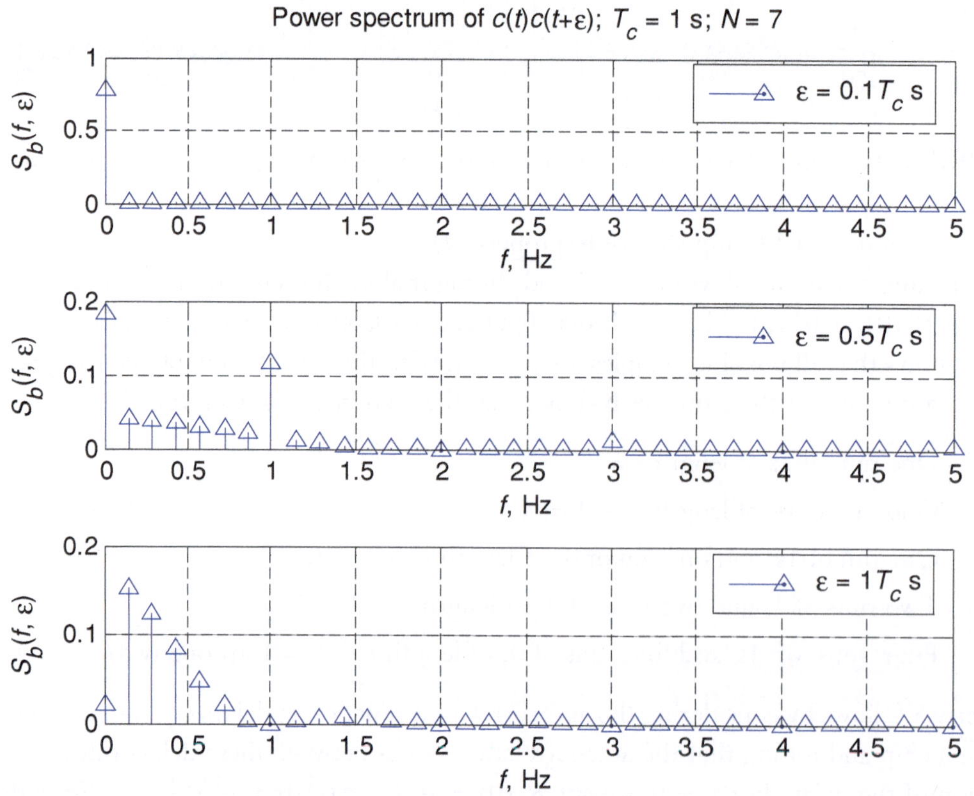

FIGURE 8: The power spectrum of $b(t, \varepsilon) = c(t)c(t + \varepsilon)$.

4.2 Gold Codes [1, 7, 8]

In communication systems with multiple users, a given user can access the system in a number of different ways among which are by being assigned a unique portion of the frequency space (frequency division multiple access, or FDMA), by being assigned a unique time portion of the signaling time frame (time division multiple access, or TDMA), or by being assigned a unique spreading code in a spread spectrum system (code division multiple access, or CDMA). In CDMA systems, it is often important that codes assigned to different users have low cross correlation with each other independent of the relative delays. Such situations are called nonsynchronous and result when the different users are at different distances from a receiver being accessed by one or more of them. Gold codes are codes whose possible cross correlations are limited to three values, given by

$$-t(n)/N, \ -1/N, \ [t(n)-2]/N, \tag{4.4}$$

where

$$t(n) = \begin{cases} 1 + 2^{0.5(n+1)} & \text{for } n \text{ odd} \\ 1 + 2^{0.5(n+2)} & \text{for } n \text{ even,} \end{cases}$$

with the code period being $N = 2^n - 1$. Gold codes are generated by modulo-2 adding certain pairs of m-sequences, known as preferred pairs, delayed relative to each other which have these cross-correlation values as well. Thus, in order to generate a family of Gold codes, it is necessary to find a preferred pair of m-sequences. The following conditions are sufficient to define a preferred pair, $\mathbf{b}$ and $\mathbf{b}'$, of m-sequences:

1. $n \neq 0 \bmod 4$; that is, n is odd or $n = 2 \bmod 4$.
2. $\mathbf{b}' = \mathbf{b}[q]$,where q is odd and either

$$q = 2^k + 1 \quad \text{or} \quad q = 2_2^k - 2^k + 1, \tag{4.5}$$

   where $\mathbf{b}[q]$ is the qth decimation of $\mathbf{b}$.

3. $\gcd(n, k) = \begin{cases} 1, & \text{for } n \text{ odd} \\ 2, & \text{for } n = 2 \bmod 4. \end{cases}$

In Item 2 above, $\mathbf{b}' = \mathbf{b}[q]$ is known as a *proper* decimation of $\mathbf{b}$ which is obtained by sampling every qth symbol of $\mathbf{b}$ and obtaining another m-sequence (which may not always be the case, thus giving an *improper* decimation).

Example 4. The m-sequence

$$\mathbf{b} = 10101\ 11011\ 00011\ 11100\ 11010\ 010000$$

when sampled every third symbol results in

$$\mathbf{b}' = 10110\ 10100\ 01110\ 11111\ 00100\ 11000\ 0$$

which is proper (spaces for ease of reading). The first condition is satisfied since $n = 5 = 1 \mod 4$. The second condition is satisfied as well, since $q = 3$ is odd and $q = 2^1 + 1$. Finally, $\gcd(5,\ 1) = 1$. Thus, a preferred pair of m-sequences has been found. A tedious manual calculation shows that for any relative shift between $\mathbf{b}$ and $\mathbf{b}'$ one of the following cross-correlation values is obtained: −9/31, −1/31, and 7/31.

Once a preferred pair of m-sequences has been found, the family of Gold codes is given by $\{b(D), b'(D), b(D) + b'(D), b(D) + Db'(D), b(D) + D^2b'(D), \ldots, b(D) + D^{N-1}b'(D)\}$. Any pair of codes from this family has the same cross-correlation values as the preferred pair of m-sequences from which they were generated. In fact, the family of Gold codes corresponding to the preferred pair of Example 3 can be generated by using different initial loads of the shift registers of Fig. 9.

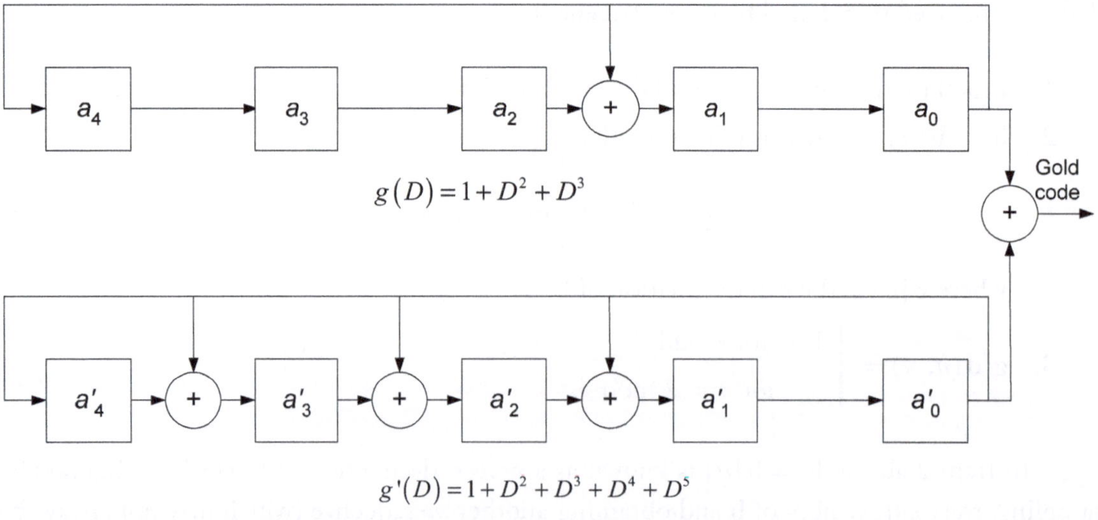

FIGURE 9: Circuit for generation of a family of Gold codes of length 31.

Several Gold codes corresponding to Example 4 and their sample cross-correlation values are given below:

–**b** and **b**′above give $C(0) = -1$.
–**b** and $D\mathbf{b}'$: $\mathbf{b} = 1010111011000111110011010010000$
$D\mathbf{b}' = 0101101010001110111110010011000$—cross correlation $= -1$.
–**b** and $D^2\mathbf{b}'$: $\mathbf{b} = 1010111011000111110011010010000$
$D^2\mathbf{b}' = 0010110101000111011111001001100$—cross correlation $= 7$.
–**b** and $D^7\mathbf{b}'$: $\mathbf{b} = 1010111011000111110011010010000$
$D^7\mathbf{b}' = 1011010100011101111100100110000$—cross correlation $= -9$.

4.3 Kasami Sequences (Small Set) [7, 8]

Consider $r = 2\nu$, where ν is an integer and let $d = 2^\nu + 1$. Let **b** be an m-sequence and let $\mathbf{b}'$ be obtained by sampling every dth symbol of **b** where $\mathbf{b}' \neq \mathbf{0}$. Then the small set of Kasami sequences is $\{\mathbf{b},\ \mathbf{b}+\mathbf{b}',\ \mathbf{b}+D\mathbf{b}',\ldots,\mathbf{b}+D^\alpha\mathbf{b}'\}$, where $\alpha = 2^\nu - 2$. These 2^ν sequences, known as the small set of Kasami sequences, have period $2^r - 1$ and have maximum magnitude cross correlation $(1 + 2^\nu)/N$.

Example 5. Consider the degree 4 entry in Table 3, which is $[2\ 3]_8 = [0\,1\,0\,0\,1\,1]_2$. Using the shift register configuration of Fig. 5(b), one period of the generated m-sequence is 100010011010111 for an initial load of 0001. For this sequence, we have $r = 4 = 2\nu$ or $\nu = 2$ and $d = 2^2 + 1 = 5$. Sampling every 5th symbol of **b** results in the sequence $\mathbf{b}' = 101101101101101$. The four Kasami sequences thereby generated are

$$\begin{aligned}\mathbf{b} &= 100010011010111\\ \mathbf{b}+\mathbf{b}' &= 001111110111010\\ \mathbf{b}+D\mathbf{b}' &= 010100101100001\\ \mathbf{b}+D^2\mathbf{b}' &= 111001000001100.\end{aligned}$$

A check of cross-correlation values results in -5/15 and 3/15, which obey the bound of $(1 + 2^\nu)/N = 5/15$.

4.4 Quaternary Sequences [9, 10]

Pseudo-random sequences other than binary-valued sequences may be useful in spread spectrum systems for several reasons. For example, four-phase spreading is used in certain spread spectrum systems by implementing two biphase systems in parallel. Use of a quaternary code would simplify such a transmitter. Another reason for quaternary-valued codes is that such codes might be found to exhibit better correlation properties than binary codes.

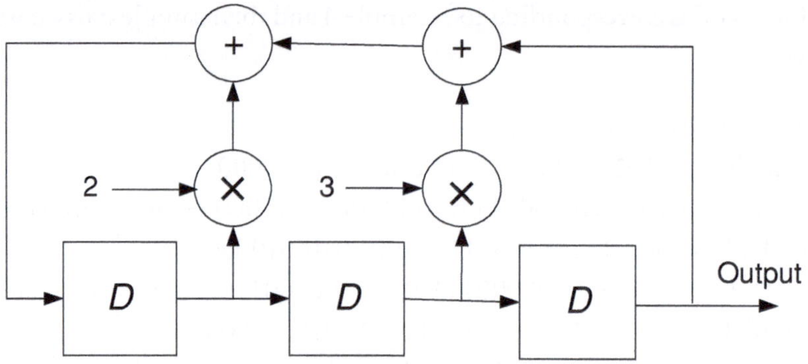

FIGURE 10: Generator for quaternary sequences of length 7.

A multiphase code family, known as the S-series, has been studied by several investigators [7, 8]. An example quaternary code generator is shown in Fig. 10. It is of interest to consider the aperiodic correlation properties of any code used for spreading purposes. The aperiodic correlation magnitudes take into account that when two sequences overlap with nonzero delay the overlap of the second sequence into the periodic extension of the first sequence may not match up in terms of phase due to the data modulation. There are three series of code families in the S-series whose properties have been studied. We limit our attention here to the $S(0)$ series. The code lengths for the $S(0)$, $S(1)$, and $S(2)$ families are all $N = 2^r - 1$, r an integer. The size of the $S(0)$ family is $N + 2$, the size of the $S(1)$ family is $\geq N^2 + 3N + 2$, and the size of the $S(2)$ family is $\geq N^3 + 4N^2 + 5N + 2$. We exhibit the maximum of the aperiodic correlation magnitude for the $S(0)$ family normalized by the code length (peak autocorrelation value) in Table 4 and a feedback generator (modulo-4 arithmetic) for an $N = 7$ code with a generator flow diagram shown in Fig. 10. The $N + 2 = 9$ possible sequences are given in Table 5.

4.5 Complementary Code Keying [6]

A quaternary code set defined in the IEEE 802.11 standard is referred to as complementary code keying (CCK). They are codes having elements a_j from the set $\{1, -1, j, -j\}$, which means that the transmitted signal is spread by phase shifts that can take on the values $\{0, \pi, \pi/2, -\pi/2\}$ radians. In fact, for the IEEE 802.11b standard, the CCK spreading phase values are chosen from the set

$$C = \{c_1, c_2, c_3, c_4, c_5, c_6, c_7, c_8\} = \left\{ \begin{array}{l} e^{j(\phi_1+\phi_2+\phi_3+\phi_4)}, e^{j(\phi_1+\phi_3+\phi_4)}, e^{j(\phi_1+\phi_2+\phi_4)}, \\ -e^{j(\phi_1+\phi_4)}, e^{j(\phi_1+\phi_2+\phi_3)}, e^{j(\phi_1+\phi_3)}, -e^{j(\phi_1+\phi_2)}, e^{j\phi_1} \end{array} \right\}. \tag{4.6}$$

TABLE 4: Worst-Case Correlation Magnitude, for the S(0) Family [10]

r	N	MAXIMUM ABS. CORRELATION	NORMALIZED CORRELATION
3	7	5.39	0.770
4	15	9.43	0.629
5	31	14.32	0.462
6	63	23.35	0.371
7	127	35.34	0.278
8	255	52.47	0.206
9	511	77.62	0.152

TABLE 5: Initial Loads and Sequences for the S(0) Family of Length 7

INITIAL LOAD	SEQUENCE
001	1001231
010	0103332
003	3003213
012	2101310
020	0202220
021	1203011
031	1302303
112	2113221
133	3312232

For the 11 Mbps data rate, each symbol represents eight bits of information. At the 5.5 Mbps data rate, four bits per symbol are transmitted. For 5.5 Mbps, CCK is used to encode four data bits (d_0 to d_3) per symbol onto the eight-chip spreading code. Data bits d_0 and d_1

TABLE 6: Encoding Table for First Two Bits (First Dibit) at Both 5.5 and 11 Mbps Data Rates

d_0, d_1	EVEN SYMBOLS	ODD SYMBOLS
00	0	π
01	$\pi/2$	$3\pi/2$
11	π	0
10	$3\pi/2$	$\pi/2$

TABLE 7: CCK Encoding Table for 5.5 Mbps Data Rate

d_2, d_3	c_1	c_2	c_3	c_4	c_5	c_6	c_7	c_8
00	j	1	j	−1	j	1	$-j$	1
01	$-j$	−1	$-j$	1	j	1	$-j$	1
11	$-j$	1	$-j$	−1	$-j$	1	j	1
10	j	−1	j	1	$-j$	1	j	1

are used to encode the ϕ_1 term above according to Table 6. Note that the phase shifts for even and odd symbols amount to giving every other symbol and extra π - radian rotation. This procedure provides nearly orthogonal codes, significantly improving the resistance to multipath and interference.

For operation at 5.5 Mbps, data bits d_2 and d_3 encode the basic symbol as shown in Table 7.

For 11 Mbps operation, the first two bits are encoded as for 5.5 Mbps operation. The remaining bits modulate the ϕ_2 through ϕ_4 phases as given in Table 8. The overall result is to modulate eight data bits onto each eight-chip spreading code.

4.6 Walsh–Hadamard Sequences [7, 8]

Walsh codes are used in second- and third-generation cellular radio systems for providing channelization, i.e., giving each user their unique piece of the communications resource. Walsh codes are orthogonal sets of 2^n binary sequences, each of length 2^n. They are defined as follows

TABLE 8: Bit-to-Chip Encoding for 11 Mbps Data Rate:
d_2, d_3 – ϕ_2 value
d_4, d_5 – ϕ_3 value
d_6, d_7 – ϕ_4 value

d_i, d_{i+1}	PHASE VALUE
00	0
01	$\pi/2$
11	π
10	$3\pi/2$

(the over bar denotes complementation):

$$\mathbf{W}_{2^1} = \begin{bmatrix} 0 & 0 \\ 0 & 1 \end{bmatrix} \equiv \begin{bmatrix} w_0 \\ w_1 \end{bmatrix}; \qquad \mathbf{W}_{2^n} = \begin{bmatrix} \mathbf{W}_{2^{n-1}} & \mathbf{W}_{2^{n-1}} \\ \mathbf{W}_{2^{n-1}} & \overline{\mathbf{W}}_{2^{n-1}} \end{bmatrix} \equiv \begin{bmatrix} w_0 \\ \vdots \\ w_{2^n-1} \end{bmatrix}. \tag{4.7}$$

For example, the Walsh set of length 4 is

$$\mathbf{W}_{2^2} = \begin{bmatrix} \mathbf{W}_2 & \mathbf{W}_2 \\ \mathbf{W}_2 & \overline{\mathbf{W}}_2 \end{bmatrix} = \begin{bmatrix} 0 & 0 & 0 & 0 \\ 0 & 1 & 0 & 1 \\ 0 & 0 & 1 & 1 \\ 0 & 1 & 1 & 0 \end{bmatrix} \tag{4.8}$$

and the Walsh set of length 8 is

$$\mathbf{W}_{2^3} = \begin{bmatrix} 0 & 0 & 0 & 0 & 0 & 0 & 0 & 0 \\ 0 & 1 & 0 & 1 & 0 & 1 & 0 & 1 \\ 0 & 0 & 1 & 1 & 0 & 0 & 1 & 1 \\ 0 & 1 & 1 & 0 & 0 & 1 & 1 & 0 \\ 0 & 0 & 0 & 0 & 1 & 1 & 1 & 1 \\ 0 & 1 & 0 & 1 & 1 & 0 & 1 & 0 \\ 0 & 0 & 1 & 1 & 1 & 1 & 0 & 0 \\ 0 & 1 & 1 & 0 & 1 & 0 & 0 & 1 \end{bmatrix}. \tag{4.9}$$

Note that the orthogonality of a pair of codes holds only if the codes are aligned.

4.7 Summary

Spreading codes are important ingredients in spread spectrum communications systems. Their ideal characteristics are that they should be easy to generate and have good auto- and cross-correlation properties. Good autocorrelation means a well-defined zero-delay peak with low nonzero-delay side lobes. Good cross-correlation properties mean cross-correlation values of low magnitude, no matter what the delay.

5 CODE ACQUISITION AND TRACKING [1]

Before data demodulation and detection can be accomplished in a spread spectrum system, the spreading code must be generated at the receiver (called the local code) and aligned with the received spreading code accounting for delay induced by the channel. The process of code alignment at the receiver is typically accomplished in two steps: alignment of the local code with the received code to within a fraction of a chip (say 1/10 chip), which is called code acquisition; tracking of the local code with the received code to within a small fraction of a chip (say 1/10 chip or less). There are two main code acquisition techniques: (1) serial search, and (2) matched filter.

For the former, i.e., serial search, an arbitrary starting point is selected in the local code, a trial correlation with the incoming code is performed, the result of this correlation is compared with a threshold, and if the threshold is exceeded, demodulation of the received spread spectrum signal is attempted. If the attempted demodulation fails, or if the threshold was not exceeded by the trial correlation, the local code is delayed a fraction of a chip (typically ½ chip), and the process is repeated. This is continued until the tracking of the incoming code by the local code is successful.

For the latter, i.e., matched filter, the magnitude of the output of a filter matched to the spreading code is compared with a threshold. When the threshold is exceeded, it is presumed that this is the delay for which the local and incoming codes are synchronous and the resulting delay is used in the demodulation of the data.

There are advantages and disadvantages to these two techniques. Two main observations are as follows:

- for long codes, serial search is substantially slower than the matched filter method for achieving acquisition;
- the complexity of the construction of the matched filter for matched filter acquisition grows substantially with the length of the spreading code.

We will first overview serial search acquisition followed by a discussion of matched filter acquisition. At the end of these discussions, we will briefly consider code tracking.

5.1 Serial Search Code Acquisition

The basic block diagram of a serial search code acquisition system is shown in Fig. 11. For simplicity, we limit our attention to acquisition in DSSS for now. The input from the dispreading mixer (multiplier) may be represented as

$$s\left(t\right)=Ad\left(t-t_{d}\right)c\left(t-t_{d}\right)c\left(t-\tau\right)\cos\left[2\pi\left(f_{\mathrm{IF}}+\Delta f\right)t+\theta\right], \tag{5.1}$$

where

A = signal amplitude at the despreading mixer output,
$d(t)$ = binary data sequence,
$c\,(t)$ = spreading code for channel of interest,
t_d = delay by the channel,
τ = delay of local code,
f_{IF} = intermediate frequency of the receiver,
Δf = frequency error introduced in the transmission (e.g., Doppler shift),
θ = unknown (as yet) phase due to channel delay, etc.

It is important to note that code acquisition and de-spreading take place before carrier acquisition or data demodulation because this allows the benefits of spread spectrum to be realized, in particular, resistance to interference and multipath and discrimination against other users. Also, it is assumed that any frequency error (e.g., due to Doppler shift) is small compared with the signal bandwidth. The bandwidth of the bandpass filter on the left in Fig. 11, therefore, is close to the *modulated signal bandwidth* (i.e., not the spread signal bandwidth). Thus, for $\tau \neq t_d$ by more than ½ a chip period (see the middle figure of Fig. 8), the signal $s\,(t)$ is essentially spread and of low spectral level. Hence, little signal power is passed by the

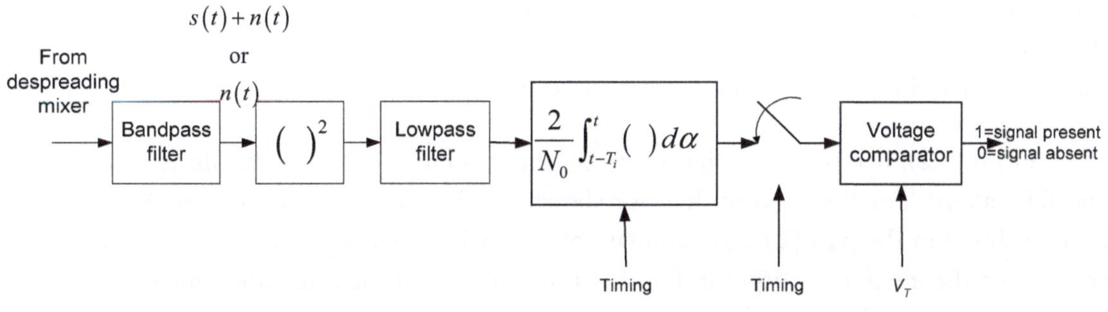

FIGURE 11: Code acquisition portion of the receiver for serial search code acquisition.

bandpass filter, little signal energy results from the integration, and the output of the integrator most likely will not cross the threshold, V_T (assuming that it is chosen properly). On the other hand, for $\tau \approx t_d$ within 1/2 a chip period (see the top figure of Fig. 8), the signal $s(t)$ is mostly de-spread and of high spectral level and of bandwidth approximately equal to the modulated signal bandwidth (as opposed to the spread signal bandwidth). Significant signal energy results from the integration, and the output of the integrator will, with high probability, cross the threshold. This alerts the tracking part of the receiver (not shown) to take over and try tracking the local code. Once tracking is established, dispreading takes place and the data is detected.

If the threshold crossing resulted from noise or a spurious correlation, the receiver must return to the code-stepping mode and continue the search for the proper alignment of the local and received codes. Clearly, the time to achieve code synchronization is a random variable. The mean and variance of this random synchronization time can be shown, respectively, to be [1]

$$\begin{aligned} \overline{T_s} &= (C-1)\, T_{\text{da}} \left(\frac{2-P_d}{2P_d}\right) + \frac{T_i}{P_d}, \\ \sigma_{T_s}^2 &\approx T_{\text{da}}^2 C^2 \left(\frac{1}{12} - \frac{1}{P_d} + \frac{1}{P_d^2}\right), \end{aligned} \tag{5.2}$$

where

$\overline{T_s}$ = mean time to acquire proper synchronization of the code,
$\sigma_{T_s}^2$ = variance of the time to acquire synchronization,
C = code uncertainty region (number of cells to be searched),
P_d = probability of detection,
P_{fa} = probability of false alarm,
T_i = integration time (time to evaluate one cell),
$T_{\text{da}} = T_i + T_{\text{fa}}\, P_{\text{fa}}$,
T_{fa} = time required to reject an incorrect phase cell.

From (5.2), it is apparent that we must obtain values for the probabilities of detection and false alarm. For the form of detector shown in Fig. 11, this is an old problem that has been analyzed in the past [11]. A summary of Urkowitz's analysis is given in [1], where it is shown that the integrator output in Fig. 11, V, at the end of the integration interval is closely approximated by a chi-squared random variable. It is a central chi-square random variable if noise alone is present at the input (i.e., codes misaligned), and noncentral chi-squared if signal plus noise is present at the input (i.e., codes aligned). These two probability density functions are given, respectively, by

$$
\begin{aligned}
p_{\text{cent chi sq}}(\alpha) &= \frac{\alpha^{(n/2)-1}}{2^{n/2}\Gamma(n/2)} \exp(-\alpha/2), \quad \alpha \geq 0, \\
p_{\text{noncent chi sq}}(\alpha) &= \frac{1}{2}\left(\frac{\alpha}{\lambda}\right)^{(n-2)/4} \exp(-\lambda/2 - \alpha/2)\, I_{(n/2)-1}\left(\sqrt{\lambda\alpha}\right), \quad \alpha \geq 0,
\end{aligned}
\tag{5.3}
$$

where

$$
\begin{aligned}
n &= 2BT_i, \\
\lambda &= n\tfrac{P}{N_0 B}, \\
P &= \text{signal power}, \\
N_0 &= \text{single-sided noise power spectral density, and} \\
I_N(\cdot) &= \text{modified Bessel function of first kind and order } N.
\end{aligned}
$$

The values of the probabilities of false alarm and detection required for computing (5.2) are given in terms of (5.3) by

$$
\begin{aligned}
P_{\text{fa}} &= \int_{V_T}^{\infty} p_{\text{cent chi sq}}(\alpha)\, d\alpha = \int_{V_T}^{\infty} \tfrac{\alpha^{(n/2)-1}}{2^{n/2}\Gamma(n/2)} \exp(-\alpha/2)\, d\alpha, \\
P_d &= \int_{V_T}^{\infty} p_{\text{noncent chi sq}}(\alpha)\, d\alpha = \int_{V_T}^{\infty} \tfrac{1}{2}\left(\tfrac{\alpha}{\lambda}\right)^{(n-2)/4} \exp(-\lambda/2 - \alpha/2)\, I_{(n/2)-1}\left(\sqrt{\lambda\alpha}\right) d\alpha,
\end{aligned}
\tag{5.4}
$$

respectively.

For computational purposes with MATLAB, these can be expressed in terms of Marcum's Q-function, which is defined as

$$
Q_M(\alpha, \beta) = \frac{1}{\alpha^{M-1}} \int_{\beta}^{\infty} x^M \exp\left(-\frac{x^2+\alpha^2}{2}\right) I_{M-1}(\alpha x)\, dx. \tag{5.5}
$$

A transformation of variables in (5.4) gives

$$
\begin{aligned}
P_{\text{fa}} &= Q_{n/2}\left(0, \sqrt{V_T}\right), \\
P_d &= Q_{n/2}\left(\sqrt{\lambda}, \sqrt{V_T}\right).
\end{aligned}
\tag{5.6}
$$

For $n = 2BT_i \gg 1$, the output of the integrator of Fig. 11 at the sampling time can be approximated as Gaussian with mean and variance given by

$$
\begin{aligned}
m_V &= n\left(\tfrac{P}{N_0 B} + 1\right), \\
\sigma_V^2 &= 2n\left(\tfrac{2P}{N_0 B} + 1\right),
\end{aligned}
\tag{5.7}
$$

respectively.

TABLE 9: Threshold Values with Accompanying Probabilities of False Alarm and Detection

V_T	P_{fa}	P_d
50.0000	0.1336	0.9327
55.0000	0.0575	0.8589
60.0000	0.0219	0.7509
65.0000	0.0075	0.6176
70.0000	0.0023	0.4755
75.0000	0.0007	0.3421
80.0000	0.0002	0.2301

Table 9 shows the probabilities of false alarm and detection versus threshold for $n = 40$ and $\lambda = 30$. It is seen that as the threshold increases, both P_{fa} and P_d decrease.

Example 6. Consider a DSSS system with code clock frequency of $f_c = 3$ MHz and suppose that $10 \log_{10}(P/N_0) = 46$ dB Hz. The propagation delay uncertainty is ± 1.2 ms. Examine the variation of code synchronization time versus V_T if $BT_i = 10$. Assume that the input bandpass filter bandwidth is 24 kHz and that the false alarm penalty is $100T_i$. Is there an optimum value of V_T?

Solution: The propagation delay uncertainty gives a value for C of (2 because of the $\pm$ uncertainty and 2 because of the ½-chip steps)

$$C = 2 \times 2\left(1.2 \times 10^{-3}\,\text{s}\right)\left(3 \times 10^6\,\text{chips/s}\right) = 14{,}400,$$

where it is assumed that the code is stepped is ½-chip increments. Also, using the given false alarm penalty, we have

$$T_{da} = T_i + T_{fa}P_{fa} = T_i + 100T_iP_{fa} = T_i(1 + 100P_{fa}).$$

Thus, the first equation of (5.2) becomes

$$\overline{T_s} = \left[(C-1)(1+100P_{fa})\left(\frac{2-P_d}{2P_d}\right) + \frac{1}{P_d}\right]T_i$$
$$= \left[14{,}399(1+100P_{fa})\left(\frac{2-P_d}{2P_d}\right) + \frac{1}{P_d}\right]T_i.$$

For the given values of B and BT_i, we have $T_i = \frac{10}{24{,}000} = 417\ \mu$s. From $n = 2BT_i = 20$ and $10\log_{10}(P/N_0) = 46$ dB Hz, we have $\lambda = n\frac{P}{N_0 B} = 20 \times \frac{10^{46/10}}{24{,}000} = 33.176$. It remains to compute P_d and P_{fa} for several values of V_T and then compute

$$\overline{T_s} = \left[14{,}399\,(1 + 100P_{\mathrm{fa}})\left(\frac{2 - P_d}{2P_d}\right) + \frac{1}{P_d}\right] \times 417 \times 10^{-6}.$$

To accomplish this, we use the MATLAB program given below with the numerical values given above. There is an optimum value for average synchronization time of about 5.74 s. The corresponding values of P_d and P_{fa} are 0.002 and 0.7727, respectively, for which $V_T = 43$. Results are given in Table 10.

```
% Example 6
%
VT = 41:0.5:46;
BTi = 10;
B = 24000;
Ti = BTi/B;
nn = 2*BTi; %nn = 2*BTi; BTi = 10
P_N0_dB = 46;
P_N0B = 10^(P_N0_dB/10)/B
lambda = nn*P_N0B
Pfa = [];
Pd = [];
for n = 1:length(VT)
   VT0 = VT(n);
   Pfa(n) = marcumq(0, sqrt(VT0), nn/2);
   Pd(n) = marcumq(sqrt(lambda), sqrt(VT0), nn/2);
end
Ti = 10/24e3; Tfa = 100*Ti;
Tda = Ti + Tfa*Pfa; C = 14400;
Ts = (C-1)*Tda.*(2-Pd)./(2*Pd) + Ti./Pd;
disp('VT Pfa Pd Ts')
disp([VT; Pfa; Pd; Ts]')
```

Clearly, there is a tradeoff between correct detections and false alarms. A false alarm is particularly expensive because it generally takes the synchronization mechanism significant time to attempt tracking as the result of a false alarm and then having to recover from it. Recall

TABLE 10: Thresholds, False Alarm and Detection Probabilities, and Corresponding Average Acquisition Times

V_T	P_{fa}	P_d	T_s
41.0	0.0037	0.8215	5.9067
41.5	0.0032	0.8099	5.8246
42.0	0.0028	0.7978	5.7708
42.5	0.0024	0.7854	5.7430
43.0	0.0020	0.7727	5.7394
43.5	0.0018	0.7596	5.7586
44.0	0.0015	0.7462	5.7990
44.5	0.0013	0.7326	5.8597
45.0	0.0011	0.7186	5.9396
45.5	0.0009	0.7044	6.0380
46.0	0.0008	0.6900	6.1544

that in this example it was assumed that the attempted tracking on a false alarm was 100 times more time consuming than tracking on a correct detection.

One approach taken to minimize the expense of attempted tracking on a false alarm is *multiple-dwell detection* wherein multiple integrations are used before the tracking mode is entered and, once it is, multiple attempts may be made to determine whether the tracking mode should be continued or exited. A typical multiple-dwell detector block diagram is shown in Fig. 12.

The logic for code alignment of such a multiple-dwell detector can be represented in terms of a flow diagram as shown in Fig. 13 [1]. In Fig. 13, it is seen that three trial integrations are carried out, all of which must indicate a successful code alignment, before the tracking mode is attempted. A miss on any one of them will cause the current code phase to be rejected and a new one tried. When in the tracking mode, two separate integrations are carried out for computing the discrimination function. If the first fails in establishing track, a second is entered and only after failure to establish track is that code phase rejected and a new code phase evaluated. It is emphasized that the logic of Fig. 13 is only one possible example of a

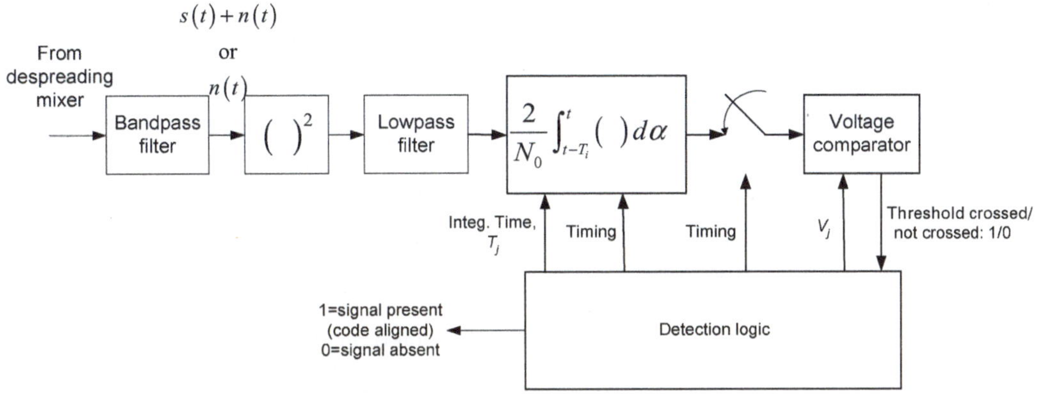

FIGURE 12: Simplified block diagram of a multiple-dwell code-alignment detector [1].

multiple-dwell code acquisition strategy. Many more possible strategies exist. The evaluation of their effectiveness is a challenging problem which will now be outlined.

An alternative way of describing the detection logic of a given multiple-dwell strategy is in terms of a state transition diagram, which shows not only the detection logic but the probabilities of transitioning from one trial integration to the next as well as the integration times. The state transition diagram corresponding to the flow diagram of Fig. 13 is shown in Fig. 14.

Each numbered circle of Fig. 14 represents a state and the arrows represent transitions between states. The labels on the arrows, where the reason for the z notation will be apparent later, give two quantities: the transition probability from one state to the next, and the time required to make that transition. For example, state 1 represents integration 1 of Fig. 13 and state 2 represents integration 2. The time required for this transition is the integration time for integrator 1, or T_1. There are two ways that the transition can occur: (1) on the basis of a threshold crossing by integration 1 on noise (codes misaligned), and (2) on the basis of a threshold crossing by integration 1 on signal plus noise (codes aligned). Thus, the probability p_{12} is a probability of false alarm in the first instance and a probability of detection in the second instance.

Consider an arbitrarily chosen path within the transition diagram, for example, starting from state 1 to 2 to 3 to 0. The product of the path labels for this series of transitions is

$$B\left(l_0, z\right) = p_{12} z^{T_1} p_{23} z^{T_2} p_{30} z^{T_3} = p_{12} p_{23} p_{30} z^{T_1+T_2+T_3}. \tag{5.8}$$

The derivative with respect to z of (5.8) gives

$$\frac{\mathrm{d}B\left(l_0, z\right)}{\mathrm{d}z} = p_{12} p_{23} p_{30} \left(T_1 + T_2 + T_3\right) z^{T_1+T_2+T_3-1}. \tag{5.9}$$

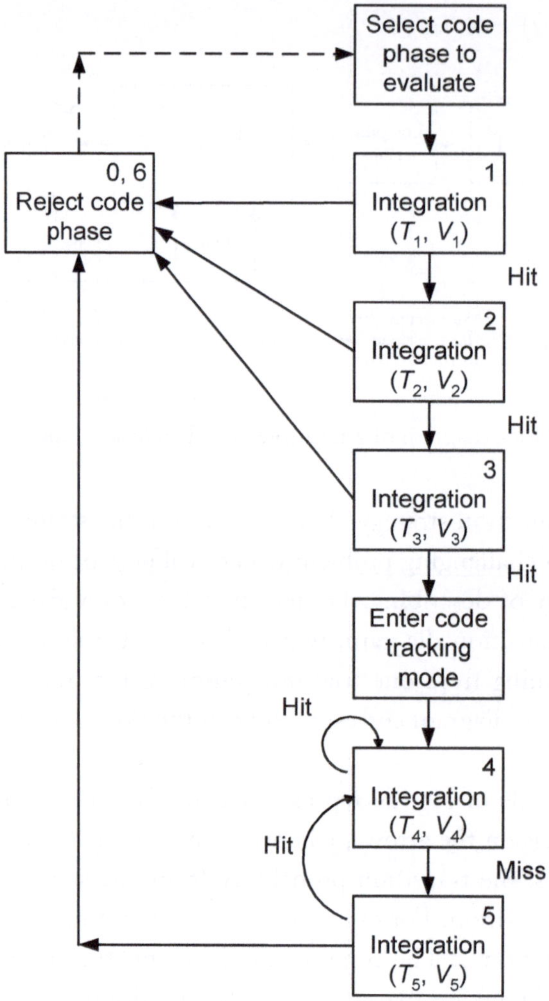

FIGURE 13: Logic flow diagram of a typical multiple-dwell detector [1].

If z is set equal to 1 in (5.9), we get

$$\Pr(l_0)\, T_l = \left[\frac{\mathrm{d}B(l_0, z)}{\mathrm{d}z}\right]_{z-1} = p_{12}p_{23}p_{30}(T_1 + T_2 + T_3)\,, \tag{5.10}$$

which is the probability of transitioning the path 1-2-3-0 times the time required to traverse the path. It is apparent that this procedure works regardless of the path chosen. Thus, if L represents the set of all paths beginning at state 1 and ending either in state 0 or state 6 of Fig. 14, we have all paths beginning at the trail of a new code phase to the rejection of that

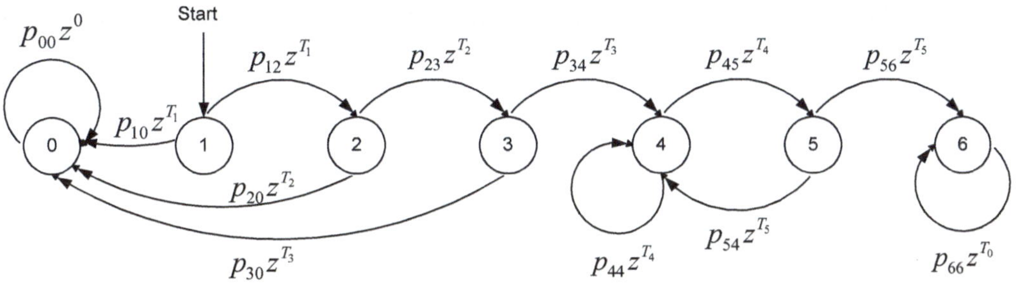

FIGURE 14: State transition diagram of the code acquisition strategy represented by the flow diagram of Figure 13 [1].

code phase. If the transition probabilities used are false alarm probabilities, it is seen that the mean time required to reject an incorrect code phase is given by

$$T_{\text{da}} = \sum_{l \in L}^{\Pr(l)} T_l = \sum_{l \in L} \left[\frac{\mathrm{d}B(l,z)}{\mathrm{d}z} \right]_{z=1}. \tag{5.11}$$

The mean time to establish track of the correct code phase can similarly be computed, except that all paths from state 1 to state 6 are considered with probabilities of detection.

The information in the state transition diagram can also be described in a state transition matrix. It can be used to compute the mean times required to accept a correct code phase or to reject an incorrect code phase. The transition matrix has rows corresponding to starting states and columns corresponding to ending states, but in a special order the reason for which will be made clear by example. Its elements are the path labels on the transition from a given row state to a given column state. For the flow graph of Fig. 13 and the state transition diagram of Fig. 14, the transition matrix is

$$\mathbf{Q}' = \begin{array}{c} \\ 0 \\ 6 \\ \\ 1 \\ 2 \\ 3 \\ 4 \\ 5 \end{array} \begin{array}{c} \begin{array}{ccccccc} 0 & 6 & 1 & 2 & 3 & 4 & 5 \end{array} \\ \left[\begin{array}{cc|ccccc} z^0 & 0 & 0 & 0 & 0 & 0 & 0 \\ 0 & z^0 & 0 & 0 & 0 & 0 & 0 \\ - & - & - & - & - & - & - \\ p_{10}z^{T_1} & 0 & 0 & p_{12}z^{T_1} & 0 & 0 & 0 \\ p_{20}z^{T_2} & 0 & 0 & 0 & p_{23}z^{T_2} & 0 & 0 \\ p_{30}z^{T_3} & 0 & 0 & 0 & 0 & p_{34}z^{T_3} & 0 \\ 0 & 0 & 0 & 0 & 0 & p_{44}z^{T_4} & p_{45}z^{T_4} \\ 0 & p_{56}z^{T_5} & 0 & 0 & 0 & p_{54}z^{T_5} & 0 \end{array}\right] \end{array} \tag{5.12}$$

where the numbers along the top row (i.e., column numbers) are to remind us of the "to" states and the numbers along the left-most column (i.e., row numbers) are to remind us of the "from" states. With this special ordering, we can identify four separate submatrices, which are defined by

$$\mathbf{Q}' = \begin{bmatrix} \mathbf{U} & \mathbf{0} \\ \mathbf{R} & \mathbf{Q} \end{bmatrix}. \tag{5.13}$$

That is, corresponding to (5.12), we identify

$$\mathbf{U} = \begin{bmatrix} 1 & 0 \\ 0 & 1 \end{bmatrix}; \quad \mathbf{0} = \begin{bmatrix} 0 & 0 & 0 & 0 & 0 \\ 0 & 0 & 0 & 0 & 0 \end{bmatrix};$$

$$\mathbf{R} = \begin{bmatrix} p_{10}z^{T_1} & 0 \\ p_{20}z^{T_2} & 0 \\ p_{30}z^{T_3} & 0 \\ 0 & 0 \\ 0 & p_{56}z^{T_5} \end{bmatrix}; \mathbf{Q} = \begin{bmatrix} 0 & p_{12}z^{T_1} & 0 & 0 & 0 \\ 0 & 0 & p_{23}z^{T_2} & 0 & 0 \\ 0 & 0 & 0 & p_{34}z^{T_3} & 0 \\ 0 & 0 & 0 & p_{44}z^{T_4} & p_{45}z^{T_4} \\ 0 & 0 & 0 & p_{54}z^{T_5} & 0 \end{bmatrix}. \tag{5.14}$$

Consider the matrix

$$\begin{aligned} \mathbf{X}_1 = \mathbf{QR} &= \begin{bmatrix} 0 & p_{12}z^{T_1} & 0 & 0 & 0 \\ 0 & 0 & p_{23}z^{T_2} & 0 & 0 \\ 0 & 0 & 0 & p_{34}z^{T_3} & 0 \\ 0 & 0 & 0 & p_{44}z^{T_4} & p_{45}z^{T_4} \\ 0 & 0 & 0 & p_{54}z^{T_5} & 0 \end{bmatrix} \begin{bmatrix} p_{10}z^{T_1} & 0 \\ p_{20}z^{T_2} & 0 \\ p_{30}z^{T_3} & 0 \\ 0 & 0 \\ 0 & p_{56}z^{T_5} \end{bmatrix} \\ &= \begin{bmatrix} p_{12}p_{20}z^{T_1+T_2} & 0 \\ p_{23}p_{30}z^{T_2+T_3} & 0 \\ 0 & 0 \\ 0 & p_{45}p_{56}z^{T_4+T_5} \\ 0 & 0 \end{bmatrix}. \end{aligned} \tag{5.15}$$

The rows of $\mathbf{X}_1$ correspond to the same states as the rows of $\mathbf{Q}$, and its columns correspond to the same states as the columns of $\mathbf{R}$. Denote the elements of $\mathbf{X}_1$ by x_{1ij}. Each term of x_{1ij} corresponds to a path of length $1 + 1 = 2$ through the state transition diagram from state i to state j. If there are no such paths then $x_{1ij} = 0$; if there is one path then there is one nonzero term; if two paths, then two nonzero terms, etc. Now if we consider $\mathbf{X}_n = \mathbf{Q}^n\mathbf{R} = \underbrace{\mathbf{QQQ}\cdots\mathbf{Q}}_{n \text{ times}}\mathbf{R}$, a similar set of statements hold except that the discussion refers to paths of length $n + 1$. From this information, we state the following as a conjecture.

Conjecture: The infinite matrix sum $\mathbf{Y} = \mathbf{R} + \mathbf{QR} + \mathbf{Q}^2\mathbf{R} + \mathbf{Q}^3\mathbf{R} + \cdots$ enumerates all paths of all lengths between inner states and end states. Specifically,

$$y_{jk} = \sum_{l \in L(j,k)} B(l, z), \tag{5.16}$$

where $L(j, \ k)$ is the set of all paths beginning at state j and ending at state k.

For the system defined by Figs. 13 and 14, the average time required to reject an incorrect cell is given by (5.11), where L denotes all paths between state 1 and states 0 or 6. Thus,

$$L = L(1, \ 0) + L(1, \ 6) \tag{5.17}$$

and (5.11) becomes

$$\begin{aligned} T_{\mathrm{da}} &= \sum_{l \in L(1,0)} \left[\frac{\mathrm{d}B(l, z)}{\mathrm{d}z}\right]_{z-1} + \sum_{l \in L(1,6)} \left[\frac{\mathrm{d}B(l, z)}{\mathrm{d}z}\right]_{z=1} \\ &= \left\{\frac{\mathrm{d}}{\mathrm{d}z} \sum_{l \in L(1,0)} B(l, z) + \frac{\mathrm{d}}{\mathrm{d}z} \sum_{l \in L(1,6)} B(l, z)\right\}_{z=1} \\ &= \left\{\frac{\mathrm{d}y_{10}}{\mathrm{d}z} + \frac{\mathrm{d}y_{16}}{\mathrm{d}z}\right\}_{z=1}. \end{aligned} \tag{5.18}$$

Since the derivative of a matrix is the matrix of derivatives, the last line of (5.18) can be obtained from $\frac{\mathrm{d}\mathbf{Y}}{\mathrm{d}z}$. Now

$$\mathbf{Y} = \mathbf{R} + \mathbf{QR} + \mathbf{Q}^2\mathbf{R} + \cdots = \left(\mathbf{I} + \mathbf{Q} + \mathbf{Q}^2 + \cdots\right)\mathbf{R} = (\mathbf{I} - \mathbf{Q})^{-1}\,\mathbf{R}. \tag{5.19}$$

We may use the matrix derivative relationship

$$\frac{\mathrm{d}}{\mathrm{d}z}\mathbf{A}^{-1} = -\mathbf{A}^{-1}\left(\frac{\mathrm{d}}{\mathrm{d}z}\mathbf{A}\right)\mathbf{A}^{-1} \tag{5.20}$$

to obtain

$$\begin{aligned} \tfrac{\mathrm{d}}{\mathrm{d}z}\mathbf{Y} &= \tfrac{\mathrm{d}}{\mathrm{d}z}(\mathbf{I} - \mathbf{Q})^{-1}\,\mathbf{R} \\ &= (\mathbf{I} - \mathbf{Q})^{-1}\left(\tfrac{\mathrm{d}}{\mathrm{d}z}\mathbf{R}\right) + \left[\tfrac{\mathrm{d}}{\mathrm{d}z}(\mathbf{I} - \mathbf{Q})^{-1}\right]\mathbf{R} \\ &= (\mathbf{I} - \mathbf{Q})^{-1}\left(\tfrac{\mathrm{d}}{\mathrm{d}z}\mathbf{R}\right) - (\mathbf{I} - \mathbf{Q})^{-1}\left(\tfrac{\mathrm{d}}{\mathrm{d}z}(\mathbf{I} - \mathbf{Q})\right)(\mathbf{I} - \mathbf{Q})^{-1}\,\mathbf{R}. \end{aligned} \tag{5.21}$$

It is convenient to define the matrix

$$\mathbf{T} = \begin{bmatrix} T_1 & & & 0 \\ & T_2 & & \\ & & \ddots & \\ 0 & & & T_n \end{bmatrix} \tag{5.22}$$

where n is the number of internal states in the state transition diagram. Thus, it is seen that

$$\left(\frac{\mathrm{d}}{\mathrm{d}z}\mathbf{R}\right)_{z=1} = (\mathbf{TR})_{z=1} \tag{5.23}$$

and

$$\left[\frac{\mathrm{d}}{\mathrm{d}z}(\mathbf{I}-\mathbf{Q})\right]_{z=1} = (-\mathbf{TQ})_{z=1}\,. \tag{5.24}$$

Using these equations in the last equation of (5.21) with $z = 1$, we obtain the following:

$$\begin{aligned}
\left(\tfrac{\mathrm{d}}{\mathrm{d}z}\mathbf{Y}\right)_{z=1} &= \left[(\mathbf{I}-\mathbf{Q})^{-1}(\mathbf{TR}) + (\mathbf{I}-\mathbf{Q})^{-1}(\mathbf{TQ})(\mathbf{I}-\mathbf{Q})^{-1}\mathbf{R}\right]_{z=1} \\
&= \left\{(\mathbf{I}-\mathbf{Q})^{-1}\mathbf{T}\left[\mathbf{R}+\mathbf{Q}(\mathbf{I}-\mathbf{Q})^{-1}\mathbf{R}\right]\right\}_{z=1} \\
&= \left\{(\mathbf{I}-\mathbf{Q})^{-1}\mathbf{T}\left[\left(\mathbf{I}+\mathbf{Q}(\mathbf{I}-\mathbf{Q})^{-1}\right)\mathbf{R}\right]\right\}_{z=1} \\
&= \left\{(\mathbf{I}-\mathbf{Q})^{-1}\mathbf{T}\left[\left(\mathbf{I}+\mathbf{Q}\left(\mathbf{I}+\mathbf{Q}+\mathbf{Q}^2+\cdots\right)\right)\mathbf{R}\right]\right\}_{z=1} \\
&= \left\{(\mathbf{I}-\mathbf{Q})^{-1}\mathbf{T}\left[\left(\mathbf{I}+\mathbf{Q}+\mathbf{Q}^2+\cdots\right)\mathbf{R}\right]\right\}_{z=1} \\
&= \left[(\mathbf{I}-\mathbf{Q})^{-1}\mathbf{T}(\mathbf{I}-\mathbf{Q})^{-1}\mathbf{R}\right]_{z=1}\,.
\end{aligned} \tag{5.25}$$

Using the result of (5.18), the average time to reject an incorrect phase cell is the sum of the elements of the first row of the final matrix of (5.25). Since incorrect phase cells are being considered, the transition probabilities are computed assuming noise only conditions. All thresholds and integration times are assumed to be known.

The probability of detection is the probability of passing from state 1 to "enter tracking mode". Arrival at the code tracking mode also implies that the system will eventually arrive at state 6. The probability of this is unity since there is no other path to an end state and the system is guaranteed to eventually reach an end state. Therefore, in the example system of Fig. 13, this

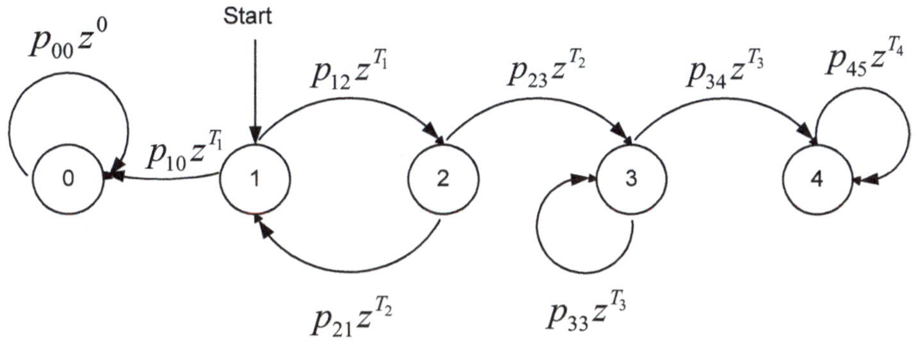

FIGURE 15: State-transition diagram for Examples 7 and 8.

probability is

$$P_d = \sum_{l \in L(1,\,6))} \Pr(l) = \sum_{l \in L(1,\,6))} B(l,\ z)_{z=1} = (y_{16})_{z=1}\,. \tag{5.26}$$

Thus, P_d is the element of the first row and second column of

$$(\mathbf{Y})_{z=1} = \left[(\mathbf{I}-\mathbf{Q})^{-1}\mathbf{R}\right]_{z=1}, \tag{5.27}$$

where the transition probabilities are now evaluated under signal plus noise conditions.

Example 7. Evaluate the matrix expressions for T_{da} and P_d for the state transition diagram of Fig. 15. Algebraic expressions are desired.

Solution: The following matrices are obtained from the state transition diagram:

$$\mathbf{R}\big|_{z=1} = \begin{bmatrix} p_{10} & 0 \\ 0 & 0 \\ 0 & p_{34} \end{bmatrix}; \qquad \mathbf{Q}\big|_{z=1} = \begin{bmatrix} 0 & p_{12} & 0 \\ p_{21} & 0 & p_{23} \\ 0 & 0 & p_{33} \end{bmatrix}. \tag{5.28}$$

From the **Q**-matrix, we obtain

$$[\mathbf{I}-\mathbf{Q}]_{z=1} = \begin{bmatrix} 1 & -p_{12} & 0 \\ -p_{21} & 1 & -p_{23} \\ 0 & 0 & 1-p_{33} \end{bmatrix} \tag{5.29}$$

from which

$$\left[(\mathbf{I}-\mathbf{Q})^{-1}\right]_{z=1} = \frac{1}{1-p_{12}p_{21}} \begin{bmatrix} 1 & p_{12} & \dfrac{p_{12}p_{23}}{1-p_{33}} \\ p_{21} & 1 & \dfrac{p_{23}}{1-p_{33}} \\ 0 & 0 & \dfrac{1-p_{12}p_{21}}{1-p_{33}} \end{bmatrix}. \tag{5.30}$$

Define the $\mathbf{T}$-matrix as a diagonal matrix with T_1, T_2, T_3 as main diagonal elements. Then

$$\begin{aligned}
\left.\frac{\mathrm{d}\mathbf{Y}}{\mathrm{d}z}\right|_{z=1} &= \left[(\mathbf{I}-\mathbf{Q})^{-1}\mathbf{T}(\mathbf{I}-\mathbf{Q})^{-1}\mathbf{R}\right]_{z=1} \\
&= \frac{1}{(1-p_{12}p_{21})^2} \begin{bmatrix} 1 & p_{12} & \dfrac{p_{12}p_{23}}{1-p_{33}} \\ p_{21} & 1 & \dfrac{p_{23}}{1-p_{33}} \\ 0 & 0 & \dfrac{1-p_{12}p_{21}}{1-p_{33}} \end{bmatrix} \begin{bmatrix} T_1 & 0 & 0 \\ 0 & T_2 & 0 \\ 0 & 0 & T_3 \end{bmatrix} \\
&\quad \times \begin{bmatrix} 1 & p_{12} & \dfrac{p_{12}p_{23}}{1-p_{33}} \\ p_{21} & 1 & \dfrac{p_{23}}{1-p_{33}} \\ 0 & 0 & \dfrac{1-p_{12}p_{21}}{1-p_{33}} \end{bmatrix} \begin{bmatrix} p_{10} & 0 \\ 0 & 0 \\ 0 & p_{34} \end{bmatrix} \\
&= \frac{1}{(1-p_{12}p_{21})^2} \\
&\quad \times \begin{bmatrix} p_{10}(p_{12}p_{21}T_2+T_1) & -\dfrac{[p_{12}p_{21}T_3+p_{33}(T_1+T_2)-T_1-T_2-T_3]\,p_{12}p_{23}p_{34}}{(1-p_{33})^2} \\ p_{10}p_{21}(T_1+T_2) & -\dfrac{[p_{12}p_{21}(p_{33}T_1-T_1+T_3)+p_{33}T_2-T_2-T_3]\,p_{23}p_{34}}{(1-p_{33})^2} \\ 0 & \dfrac{(1-p_{12}p_{21})^2\,p_{34}T_3}{(1-p_{33})^2} \end{bmatrix}.
\end{aligned} \tag{5.31}$$

Taking the sum of the elements in the first row, we obtain

$$T_{\mathrm{da}} = \frac{p_{10}(p_{12}p_{21}T_2+T_1)}{(1-p_{12}p_{21})^2} - \frac{[p_{12}p_{21}T_3+p_{33}(T_1+T_2)-T_1-T_2-T_3]\,p_{12}p_{23}p_{34}}{(1-p_{12}p_{21})^2(1-p_{33})^2}. \tag{5.32}$$

This can be simplified to

$$T_{\text{da}} = \frac{p_{10}\left(T_1 + p_{12}p_{21}T_2\right)}{\left(1 - p_{12}p_{21}\right)^2} + \frac{p_{12}p_{23}p_{34}\left(T_1 + T_2\right)}{\left(1 - p_{12}p_{21}\right)^2\left(1 - p_{33}\right)} + \frac{p_{12}p_{23}p_{34}T_3}{\left(1 - p_{12}p_{21}\right)\left(1 - p_{33}\right)^2}. \tag{5.33}$$

Recall that noise-alone conditions are used to compute the various transition probabilities. To get P_d, we compute

$$\mathbf{Y}|_{z=1} = \left[(\mathbf{I} - \mathbf{Q})^{-1}\,\mathbf{R}\right]_{z=1} = \frac{1}{1 - p_{12}p_{21}}\begin{bmatrix} p_{10} & \dfrac{p_{12}p_{23}p_{34}}{1 - p_{33}} \\ p_{10}p_1 & \dfrac{p_{23}p_{34}}{1 - p_{33}} \\ 0 & \dfrac{p_{34}\left(1 - p_{12}p_{21}\right)}{1 - p_{33}} \end{bmatrix} \tag{5.34}$$

and take the element in the upper-right-hand corner as P_d. This gives

$$P_d = \frac{p_{12}p_{23}p_{34}}{\left(1 - p_{12}p_{21}\right)\left(1 - p_{33}\right)} = \frac{p_{12}p_{23}}{1 - p_{12}p_{21}};\; p_{34} = 1 - p_{33}. \tag{5.35}$$

Recall the signal plus noise conditions are used to compute the various transition probabilities.

Example 8. Assume the parameters of Example 6, namely, a code clock frequency of $f_c = 3$ MHz and $10\log_{10}(P/N_0) = 46$ dB Hz. The propagation delay uncertainty is ± 1.2 ms and the code is stepped in ½-chip steps. However, the double-dwell system of Fig. 15 is used with time–bandwidth products for integrations 1, 2, and 3 chosen as 4, 10, and 50, respectively. The bandwidth is still 24 kHz, giving integration times of $T_1 = 1.67 \times 10^{-4}$ s, $T_2 = 4.17 \times 10^{-4}$ s, and $T_3 = 2.083 \times 10^{-3}$ s. The thresholds are chosen to give probabilities of detection of $p_{12} = p_{23} = 0.9$ and $p_{33} = 0.99$. Use these to compute the corresponding thresholds, then the probabilities of false alarm, and finally the mean synchronization time.

Solution: Use of the marcumq function in MATLAB results in the following:

$BT = 4;\; V_T = 11.5000;\; P_{fa} = 0.1749 = p_{12};\; P_d = 0.8959 = p'_{12}$

$BT = 10;\; V_T = 37.0000;\; P_{fa} = 0.0117 = p_{23};\; P_d = 0.9003 = p'_{23}$

$BT = 50;\; V_T = 201.5000;\; P_{fa} = 8 \times 10^{-11} = p_{33};\; P_d = 0.9902 = p'_{33}$

$p_{10} = 1 - p_{12};\; p_{21} = 1 - p_{23};\; p_{34} = 1 - p_{33};\; p_{45} = 1$

$p'_{10} = 1 - p'_{12};\; p'_{21} = 1 - p'_{23};\; p'_{34} = 1 - p'_{33};\; p'_{45} = 1$(primed probabilities are $s + n$).

The **R** and **Q** matrices for noise-alone conditions are

$$\mathbf{R}\big|_{z=1} = \begin{bmatrix} 1-0.1749 & 0 \\ 0 & 0 \\ 0 & 1-8\times 10^{-11} \end{bmatrix} = \begin{bmatrix} 0.8251 & 0 \\ 0 & 0 \\ 0 & 1-8\times 10^{-11} \end{bmatrix};$$

$$\mathbf{Q}\big|_{z=1} = \begin{bmatrix} 0 & 0.1749 & 0 \\ 1-0.0117 & 0 & 0.0117 \\ 0 & 0 & 8\times 10^{-11} \end{bmatrix} \approx \begin{bmatrix} 0 & 0.1749 & 0 \\ 0.9883 & 0 & 0.0117 \\ 0 & 0 & 8\times 10^{-11} \end{bmatrix}.$$

Compute

$$\left.\frac{d\mathbf{Y}}{dz}\right|_{z=1} = \left[(\mathbf{I}-\mathbf{Q})^{-1}\mathbf{T}(\mathbf{I}-\mathbf{Q})^{-1}\mathbf{R}\right]_{z=1} = \begin{bmatrix} 0.0002879 & 0.0000069 \\ 0.0006953 & 0.0000371 \\ 0 & 0.0020833 \end{bmatrix}.$$

The mean time to synchronization is the sum of the elements in the first row, or $T_{da} \approx 0.000295$ s $= 295$ μs. The question is whether readjustment of the thresholds or integrations times can decrease this. Computer evaluation shows that there is an optimum set of thresholds and integration times.

For signal plus noise conditions, the **R** and **Q** matrices are

$$\mathbf{R}'\big|_{z=1} = \begin{bmatrix} 1-0.8959 & 0 \\ 0 & 0 \\ 0 & 1-0.9902 \end{bmatrix} = \begin{bmatrix} 0.1041 & 0 \\ 0 & 0 \\ 0 & 0.0098 \end{bmatrix};$$

$$\mathbf{Q}'\big|_{z=1} = \begin{bmatrix} 0 & 0.8959 & 0 \\ 1-0.9003 & 0 & 0.9003 \\ 0 & 0 & 0.9902 \end{bmatrix} \approx \begin{bmatrix} 0 & 0.8959 & 0 \\ 0.0997 & 0 & 0.9003 \\ 0 & 0 & 0.9902 \end{bmatrix}.$$

We compute the matrix

$$\mathbf{Y}\big|_{z=1} = \left[(\mathbf{I}-\mathbf{Q})^{-1}\mathbf{R}\right]_{z=1}$$

under signal plus noise conditions to find P_d as the upper-right-hand element. Carrying out the numerical calculations, we get $P_d = 0.8857$. Using signal plus noise transition probabilities in (5.35) as a check, we get

$$P_d = \frac{p'_{12}p'_{23}}{1-p'_{12}p'_{21}} = \frac{0.8959 \times 0.9003}{1-0.8959\,(1-0.9003)} = 0.8857,$$

which is exactly the same as obtained with direct matrix calculations.

5.2 Matched Filter Code Acquisition [1, 12]

A matched filter has impulse response which is the delayed time reverse of the signal to which it is matched. Thus, for a time-limited signal $s(t)$ that is zero for $t < 0$ and $t > T$, the impulse response of the filter matched to this signal is $h(t) = s(t_0 - t)$, where t_0 is the time of peak output and is usually chosen to make the filter causal. Thus, for causality, $t_0 \geq T$ in this case. For the choice of $t_0 = T$, the output of the matched filter for an arbitrary input $x(t)$ is

$$y(t) = \int_{-\infty}^{\infty} h(\tau)\, x(t-\tau)\, d\tau = \int_{0}^{T} s(T-\tau) x(t-\tau)\, d\tau, \tag{5.36}$$

which, at time $t = T$, is

$$y(T) = \int_{0}^{T} s(T-\tau) x(T-\tau)\, d\tau = \int_{0}^{T} s(\lambda) x(\lambda)\, d\lambda. \tag{5.37}$$

This is a well-known property of a matched filter—its output at the optimum sampling time is the cross-correlation between the input and the signal to which it is matched. If the input is $x(t) = s(t) + n(t)$, where $n(t)$ is the white noise of the two-sided power spectral density of $N_0/2$, the output is

$$\begin{aligned} y(T) &= \int_{0}^{T} s(\lambda)\left[s(\lambda) + n(\lambda)\right] d\lambda = \int_{0}^{T} s^2(\lambda)\, d\lambda + N \\ &= E_s + N, \end{aligned} \tag{5.38}$$

where $E_s = \int_0^T s^2(\lambda)\, d\lambda$ is the signal energy and N is a zero-mean random variable with variance $\sigma_N^2 = \frac{N_0 E_s}{2}$. Thus, the peak signal squared to mean-square noise at the output is

$$SNR_{\text{out}} = \frac{E_s^2}{(N_0 E_s/2)} = \frac{2E_s}{N_0} \tag{5.39}$$

which is a well-known property of matched filters [11].

Since a matched filter performs the function of a correlator, the correlation operation in the acquisition circuitry of a DSSS receiver may be replaced by a matched filter. A conceptual block diagram for such a receiver is shown in Fig. 16. The upper three boxes constitute the matched filter code acquisition circuitry. The remaining boxes represent the fine code tracking and data detection. The bandpass matched filter (it could be realized at quadrature baseband) is matched to K chips of the spreading code modulating the IF carrier, so its impulse response

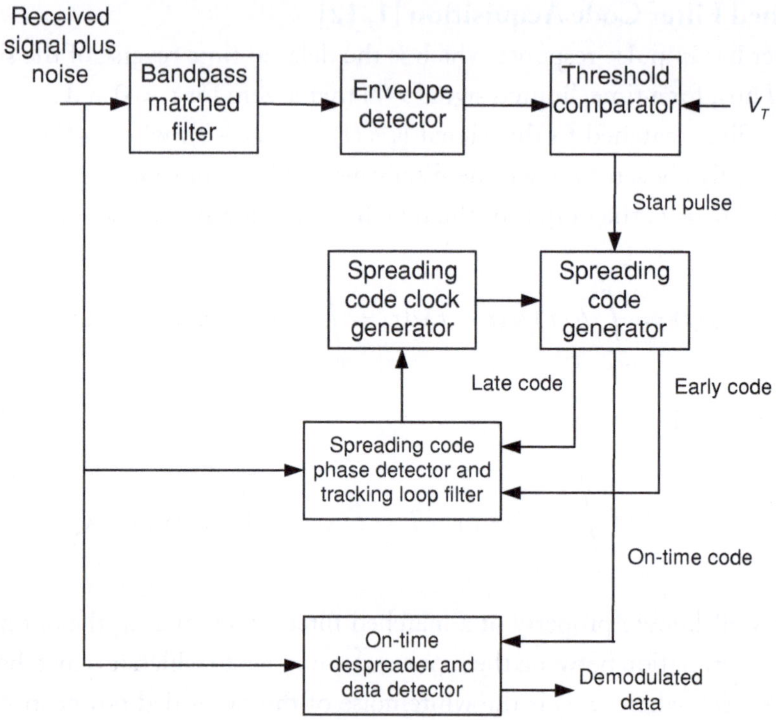

FIGURE 16: DSSS receiver using matched filter code acquisition [1].

is

$$h(t) = 2c(KT_c - t)\cos\omega_{\mathrm{IF}}t, \qquad 0 \le t \le KT_c. \tag{5.40}$$

For signal alone at the input, its output is

$$\begin{aligned}
y(t) &= \int_{-\infty}^{\infty} h(\tau)s_{\mathrm{in}}(t-\tau)\,\mathrm{d}\tau \\
&= \int_{0}^{KT_c} [2c(KT_c - \tau)\cos\omega_{\mathrm{IF}}\tau]\left\{\sqrt{2P}c(t-\tau-t_d)\cos[\omega_{\mathrm{IF}}(t-\tau)+\phi]\right\}\mathrm{d}\tau \\
&= \sqrt{2P}\int_{0}^{KT_c} c(KT_c-\tau)c(t-\tau-t_d)\{\cos(\omega_{\mathrm{IF}}t+\phi)+\cos[\omega_{\mathrm{IF}}(t-2\tau)+\phi]\}\,\mathrm{d}\tau \\
&\approx \sqrt{2P}\left[\int_{0}^{KT_c} c(KT_c-\tau)c(t-\tau-t_d)\,\mathrm{d}\tau\right]\cos(\omega_{\mathrm{IF}}t+\phi) \\
&= \sqrt{2P}R_{c,KT_c}(KT_c - t + t_d)\cos(\omega_{\mathrm{IF}}t+\phi),
\end{aligned} \tag{5.41}$$

where $R_{c,KT_c}(\lambda)$ is the code correlation function over a duration of KT_c and data modulation has been ignored for the time being for simplicity. The output of the envelope detector is the envelope or $\sqrt{2P}\left|R_{c,\,KT_c}\left(KT_c - t + t_d\right)\right|$. The time at which this function is a maximum is used as an indication of the epoch at which the local code is to be delayed for tracking of the incoming code.

If there is a frequency error between the incoming signal and the matched filter center frequency (due to Doppler shift, say), then the output is degraded by a factor dependent on the frequency error and integration time [1]. If the frequency error is zero, there is still degradation due to the correlation being over only a portion of the code duration and also because of possible data transitions during the correlation duration.

In spite of these possible degradations, matched filter code acquisition is attractive because it speeds up the average acquisition time by a factor of roughly K.

Matched filter acquisition is often implemented digitally. If N is the number of samples taken per chip, then for an initial code uncertainty of M chips it can be shown that, under the ideal conditions of $P_d = 1$ and $P_{\text{fa}} = 0$, the average acquisition time is [1]

$$\overline{T_s} \approx \frac{T_c}{N} + \frac{MT_c}{2}. \tag{5.42}$$

This is a considerable savings over serial search.

The schematic of a digitally implemented matched filter realization for DSSS code acquisition is shown in Fig. 17. Typical correlation function envelopes are shown in Fig. 18 for the 11-chip code {1 −1 1 1 −1 1 1 1 −1 −1 −1} for two samples per chip. A repetition of four transmitted bits is shown, where each bit contains a code repetition.

An exact analysis of matched filter acquisition is complex and the reader is referred to the literature [12].

Example 9. Compare serial search and digital matched filter acquisition under ideal conditions of $P_d = 1$ and $P_{\text{fa}} = 0$, an initial code uncertainty of 10,000 half-chips, and a chip rate of 1 Mchip/s. For the matched filter case, assume $N = 2$ samples/chip. Assume an integration time equivalent to 100 chips for the serial search case.

Solution: The given chip rate means that $T_c = 1\ \mu$s and the integration time is $T_i = 10^{-4}$ s. For $P_d = 1$ and $P_{\text{fa}} = 0$, (5.2) simplifies to

$$\overline{T_s} = \left(\frac{C+1}{2}\right) T_i = \frac{10{,}001}{2}\left(10^{-4}\right) = 0.50005 \text{ s}.$$

For the matched filter case $M = 10{,}000/2 = 5000$ (the matched filter steps in chip intervals) and

$$\overline{T_s} \approx \frac{T_c}{N} + \frac{MT_c}{2} = \frac{10^{-6}}{2} + \frac{(5000)\left(10^{-6}\right)}{2} = \frac{1}{2}(5001)\left(10^{-6}\right) = 0.0025 \text{ s}.$$

The advantage of matched filter acquisition in terms of acquisition time is clear.

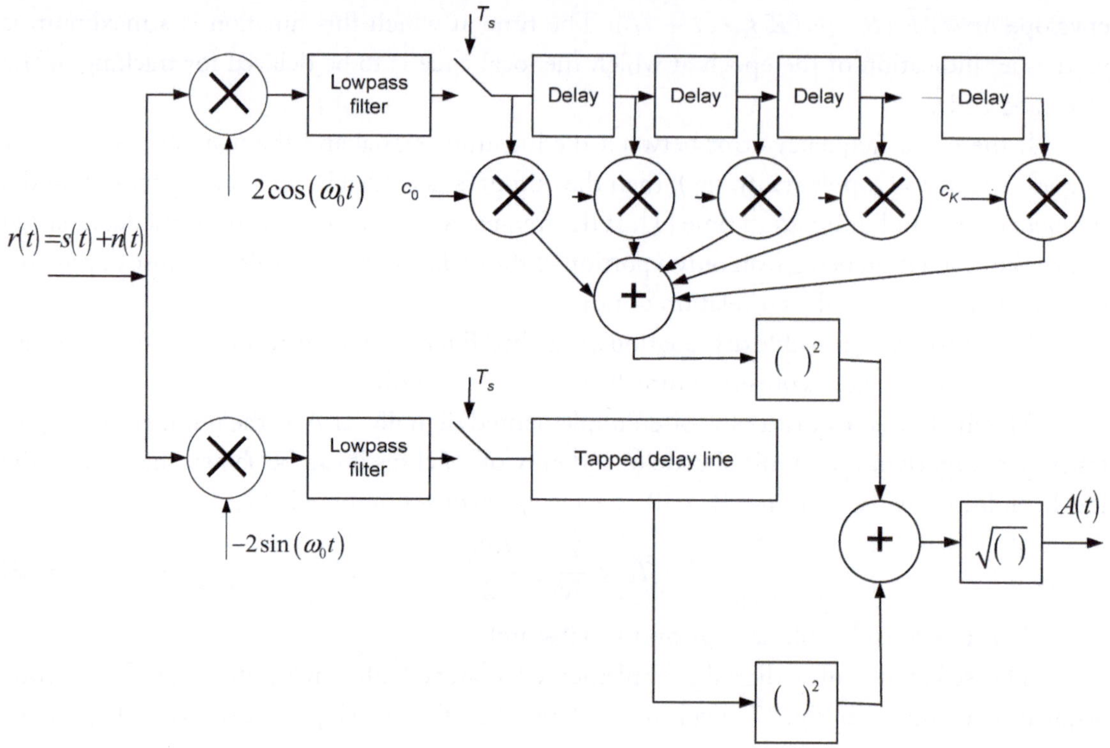

FIGURE 17: Digital implementation of a noncoherent matched filter for acquisition [1].

5.3 Tracking in Spread Spectrum

Once the locally generated code has been stepped to within a fraction of a chip (typically ½ chip) of the received code, the receiver is switched to the tracking mode and the local code is tracked to within a very small fraction of the incoming code (ideally local and received codes are coincident, but noise will cause jitter around this ideal value). To see how this might be accomplished, consider the idealized case of a received signal consisting of a spreading code plus noise (i.e., data and modulation on a carrier are being ignored for now):

$$x_r(t) = \sqrt{P}c(t - T_d) + n(t), \tag{5.43}$$

where P is the average power of the input signal component. The mechanism shown by the block diagram of Fig. 19 will track this signal if the received and local codes are within a chip of each other. To show that this is the case, consider the time average of the output of the

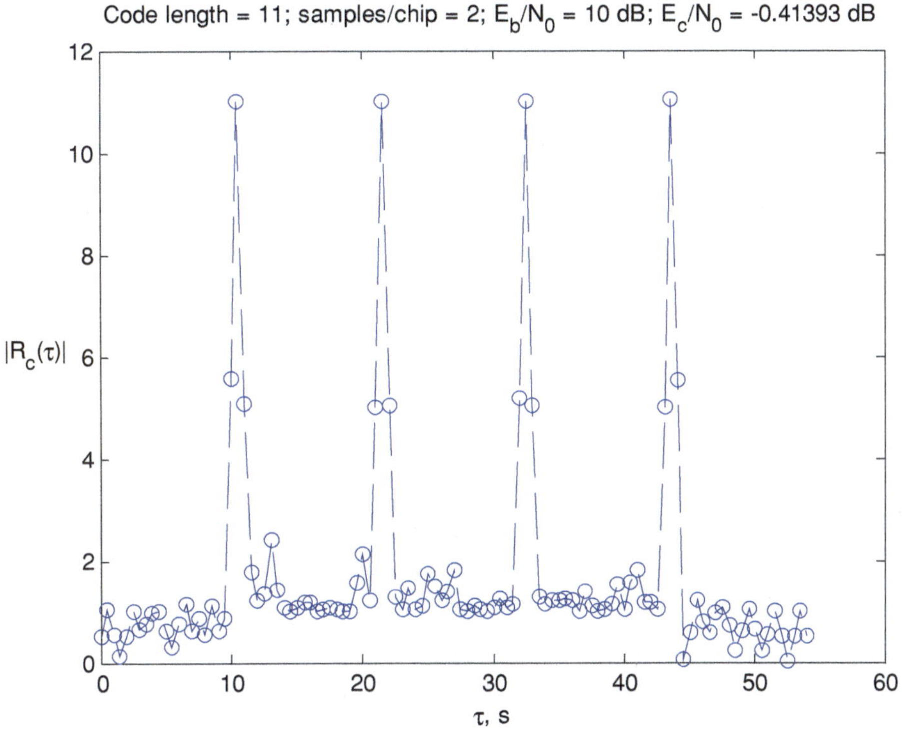

FIGURE 18: Matched filter output for sequence of four bits or four 11-chip code repetitions.

differencer, which is

$$\begin{aligned}\varepsilon\left(t,\ T_d,\ \widehat{T}_d\right) &= \left\langle K_1\sqrt{\tfrac{P}{2}}c\left(t-T_d\right)\left[c\left(t-\widehat{T}_d-\tfrac{\Delta}{2}T_c\right)-c\left(t-\widehat{T}_d+\tfrac{\Delta}{2}T_c\right)\right]\right\rangle \\ &\triangleq K_1\sqrt{\tfrac{P}{2}}D_\Delta\left(T_d,\ \widehat{T}_d\right)+n_{\text{self-noise}}\left(t\right),\end{aligned} \tag{5.44}$$

where $D_\Delta\left(T_d,\ \widehat{T}_d\right)$ is the average of (5.44) over a time interval of the order of the code duration and is given by

$$\begin{aligned}D_\Delta\left(T_d,\ \widehat{T}_d\right) &= \tfrac{1}{NT_c}\int\limits_{-NT_c/2}^{NT_c/2} c\left(t-T_d\right)\left[c\left(t-\widehat{T}_d-\tfrac{\Delta}{2}T_c\right)-c\left(t-\widehat{T}_d+\tfrac{\Delta}{2}T_c\right)\right]\mathrm{d}t \\ &\triangleq R_c\left(T_d-\widehat{T}_d-\tfrac{\Delta}{2}T\right)-R_c\left(T_d-\widehat{T}_d+\tfrac{\Delta}{2}T\right) \\ &= R_c\left[\left(\delta-\tfrac{\Delta}{2}\right)T\right]-R_c\left[\left(\delta+\tfrac{\Delta}{2}\right)T\right] \\ &\triangleq D_\Delta\left(\delta\right).\end{aligned} \tag{5.45}$$

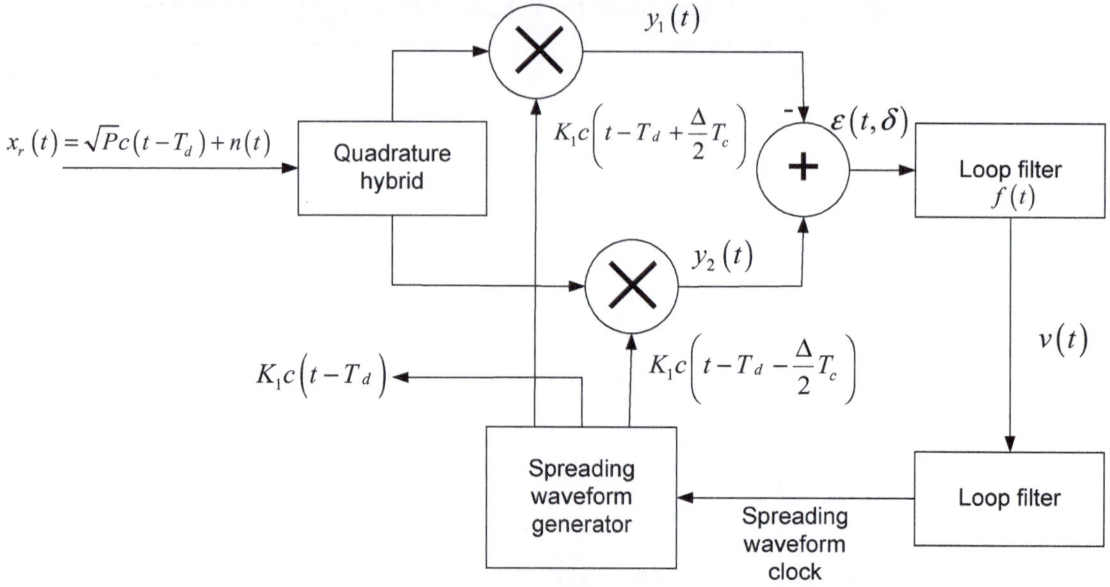

FIGURE 19: Baseband delay-lock tracking loop [1].

The second term is the AC component, referred to as the self-noise since it is the result of code products that do not aid tracking and $R_c(\tau)$ is the code correlation function. $D_\Delta(\delta)$ is plotted in Fig. 20 for several values of Δ. It is seen that any of these can serve as a suitable control signal for the voltage controlled oscillator (VCO) which provides the clock signal for driving the local code generator of Fig. 19, but the discriminator characteristics for $\Delta = 1$ and 2 are particularly attractive because of their interior linear regions. From these plots, it is apparent that if the local code lags the incoming code the discriminator characteristic will provide a signal to the VCO which speeds it up, whereas if the local code leads the incoming code the discriminator characteristic will provide a signal to the VCO which slows it down. Thus, the codes will be maintained in close synchronism which is not exactly zero due to the action of the noise at the input. The operation of this system in noise can be characterized through the application of standard phase-lock loop analysis techniques [1].

The modifications needed to make this tracking loop practical are ones to accommodate modulated signals, i.e., accommodations for data times the code times a carrier. A loop structure which allows for data on a carrier is called the noncoherent delay-lock tracking loop and is shown in block diagram form in Fig. 21. The carrier is accommodated by the inphase and quadrature-channel mixers in the upper-left-hand corner, and the presence of data is accommodated by the squarers in the upper-middle portion of the diagram. The discriminator characteristic for this

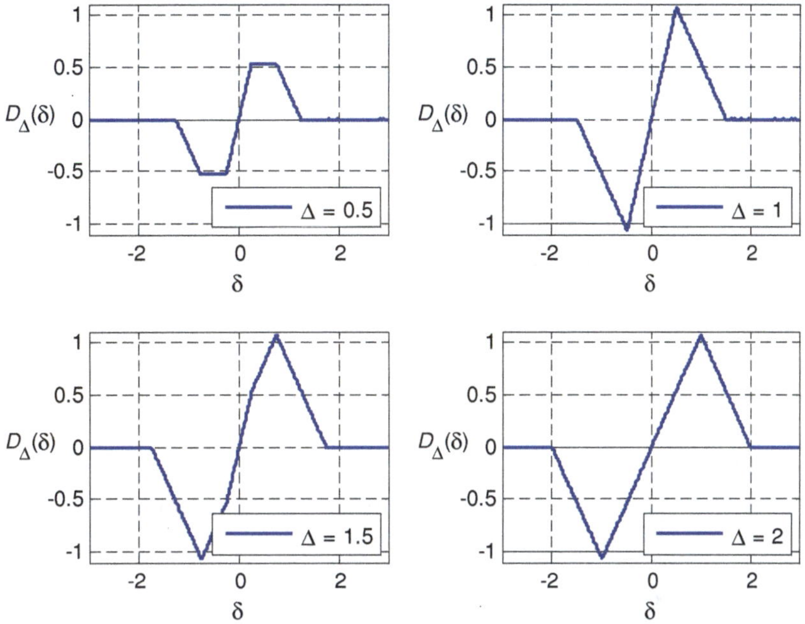

FIGURE 20: Delay-lock discriminator dc outputs for a 15-chip m-sequence for various values of Δ.

circuit is proportional to the difference of the squares of the code correlation functions delayed and advanced, respectively, by $\Delta/2$. For proper choice of Δ, they exhibit a linear interior region, making them suitable for driving the VCO in the proper direction.

The code tracking jitter variance for the noncoherent delay-lock tracking loop is given by

$$\sigma^2_{\delta,\,\mathrm{DLL}} = \frac{1}{2\rho_L}\left(1 + \frac{2}{\rho_{\mathrm{IF}}}\right), \tag{5.46}$$

where

$$\rho_L = \frac{P}{N_0 B_L} = \text{signal-to-noise ratio in the loop bandwidth}, B_L,$$

$$\rho_{\mathrm{IF}} = \frac{P}{N_0 B_{\mathrm{IF}}} = \text{signal-to-noise ratio in the receiver IF bandwidth}, B_{\mathrm{IF}}.$$

There are many variations of code tracking loops. Another important one is the tau-dither noncoherent tracking loop which requires less hardware than the delay-lock tracking loop at the expense of slightly worse tracking jitter variance. The block diagram of the tau-dither tracking loop is shown in Fig. 22. It is seen that the early and late versions of the locally generated code are time shared in the same channel by virtue of the slow switching function $q(t) = \pm 1$. This points out another advantage of the tau-dither tracking loop over the delay-lock tracking

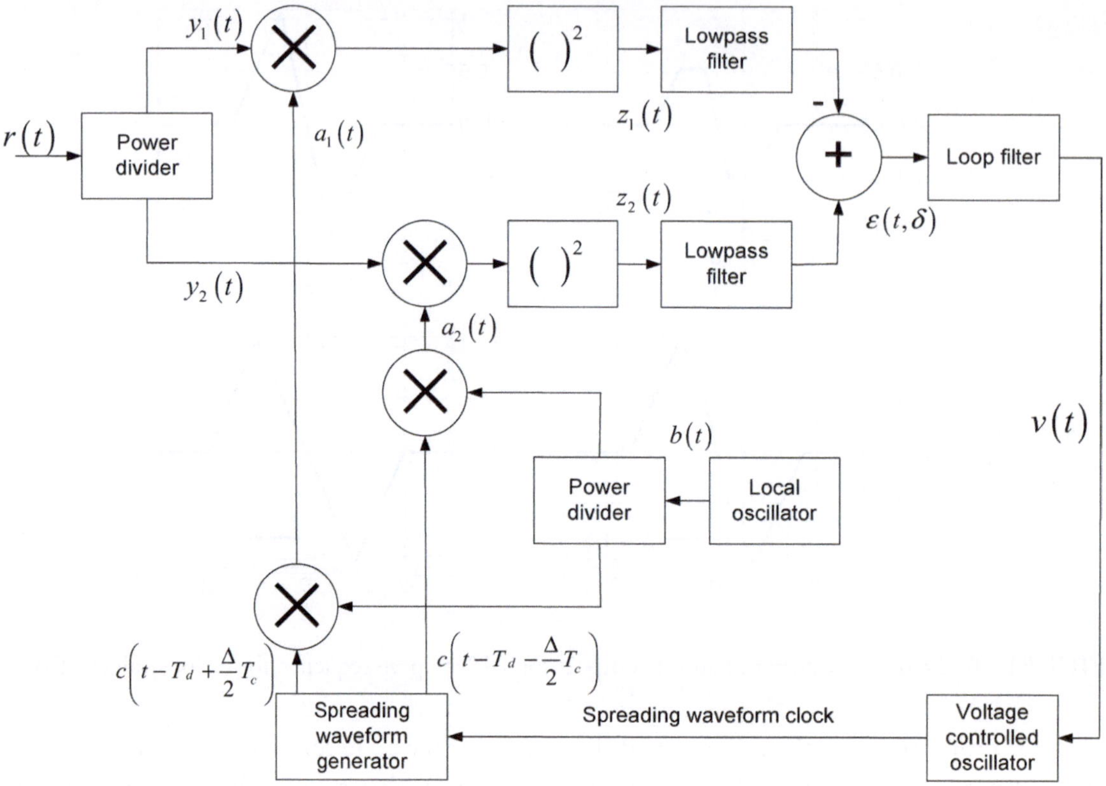

FIGURE 21: Noncoherent delay-lock code tracking loop [1].

loop—possible gain and phase imbalances between the two channels of the delay-lock tracking loop are avoided in the tau-dither loop because a single channel is time shared between the early and late codes. The tracking jitter variance of the tau-dither loop, for BPSK spreading and a switching frequency of $f_q = B_L/4$ Hz, is given by

$$\sigma^2_{\delta,\,\mathrm{TDL}} = \frac{1}{2\rho_L}\left(1.811 + \frac{3.261}{\rho_{\mathrm{IF}}}\right), \tag{5.47}$$

where ρ_L and ρ_{IF} are as defined in (5.46).

Example 10. Compare the tracking jitter standard deviations of tau-dither tracking and delay-lock tracking loops for the following parameters:

$$\rho_{\mathrm{IF}} = \frac{P}{N_0 B_{\mathrm{IF}}} = 10,$$
$$B_L = B_{\mathrm{IF}}/50.$$

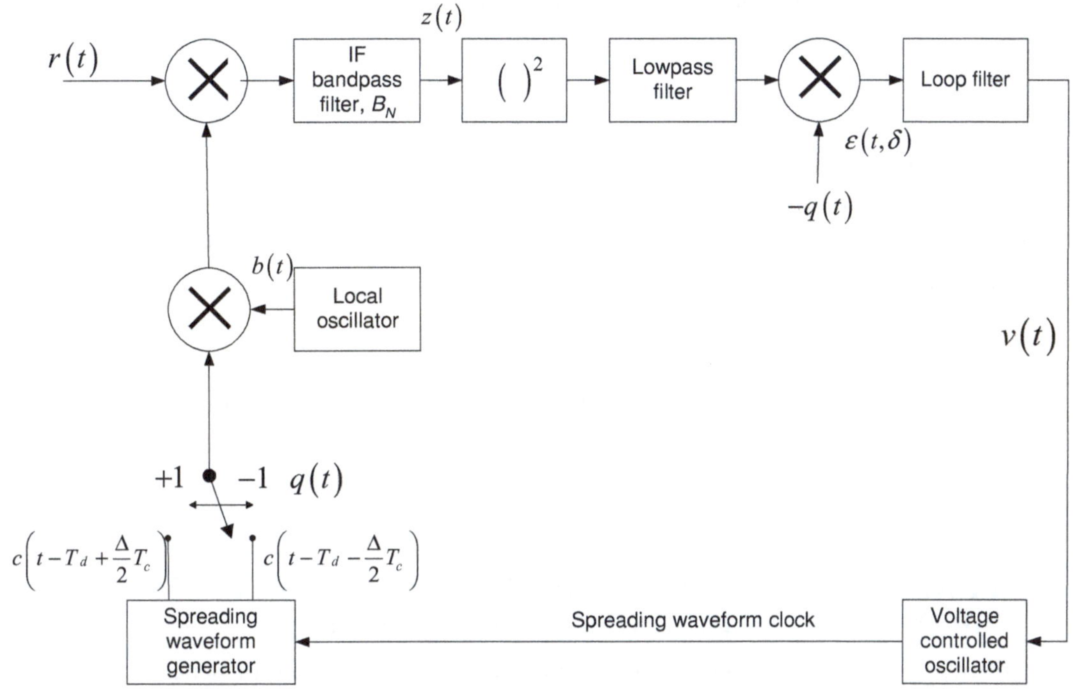

FIGURE 22: Block diagram of a tau-dither code tracking loop [1].

Solution: From the given data, we find that $\rho_L = \frac{P}{N_0 B_L} = \frac{P}{N_0 B_{IF}} \frac{B_{IF}}{B_L} = 10\,(50) = 500$. Thus,

$$\sigma^2_{\delta,\,\mathrm{TDL}} = \frac{1}{2\rho_L}\left(1.811 + \frac{3.261}{\rho_{\mathrm{IF}}}\right) = \frac{1}{2\,(500)}\left(1.811 + \frac{3.261}{10}\right) = \frac{2.1371}{2\,(500)} = 2.1371 \times 10^{-3}\mathrm{s}^2,$$

$$\sigma^2_{\delta,\mathrm{DLL}} = \frac{1}{2\rho_L}\left(1 + \frac{2}{\rho_{\mathrm{IF}}}\right) = \frac{1}{2\,(500)}\left(1 + \frac{2}{10}\right) = \frac{1.2}{2\,(500)} = 1.2 \times 10^{-3}\ \mathrm{s}^2.$$

The respective standard deviations are

$$\sigma_{\delta,\,\mathrm{TDL}} = 0.0462\ \mathrm{s},$$
$$\sigma_{\delta,\,\mathrm{DLL}} = 0.0346\ \mathrm{s}.$$

In terms of standard deviation, which gives one basis of comparison for relative performance, we see that the two tracking loops are fairly close in this particular example.

5.4 Summary

In this section, the synchronization of the local de-spreading code at the receiver with the spreading code on the received signal has been considered. Generally, this consists of two steps:

(1) initial acquisition, where the local and received codes are aligned to within 1/2 chip or less;nd (2) tracking, or fine tuning the initial alignment, to within a small fraction of a chip. The latter is typically implemented with a phase-lock-loop type of feedback structure. The former is typically implemented either as a serial search algorithm or as a matched filter-based structure. Both were discussed in this section, with more mathematical details being given for the serial search procedure than for matched filter-based structures. The reason for this is that, since longer integration times are possible with serial search, the effects of code correlation side lobes are not usually an issue, whereas they are for matched filter implementations since hardware limitations dictate correlation over shorter code segments in the matched filter case. For a given integration time, matched filter acquisition gives by far lower average synchronization times than serial search. The discussion in this chapter is centered around acquisition for DSSS. Code acquisition considerations for FHSS are similar to those for DSSS, at least mathematically, although the implementation of the hardware is decidedly different.

6 PERFORMANCE OF SPREAD SPECTRUM SYSTEMS OPERATING IN JAMMING—NO CODING

The performance of a spread spectrum communication system in the presence of AWGN is the same as the system without spread spectrum using the same data modulation technique as the spread spectrum system. In order to make a spread spectrum communication system's performance unacceptable, an enemy might resort to jamming, i.e., radiating a signal in the same band being used by the spread spectrum system in order to raise its error probability to an unacceptable level.

Another possible source of interference in spread spectrum systems is multiple-access interference. This will be considered in Section 7.

Jamming can take many forms. Some examples are

- Jamming with wideband (barrage) noise;
- Jamming with narrowband or partial band noise;
- Jamming with a single frequency;
- Jamming with a comb of (multiple) frequencies;
- Jamming with pulsed noise;
- Jamming with a repeated replica of the communicator's signal.

These are basically arranged in order of least complex to most complex. We will take up the performance analysis of each in turn except for the last.

6.1 Barrage Noise Jamming

This is the simplest jamming of all those listed, both to implement and to analyze. If the jammer is J watts and it is radiated as wideband noise, then the communication system noise spectral level is raised from N_0 W/Hz to $N_0 + J/B_{ss}$ W/Hz, where B_{ss} is the single-sided bandwidth of the spread spectrum signal. For direct sequence BPSK spreading, $B_{ss} \approx 2/T_c$ Hz, where T_c is the chip duration. Thus, the bit error probability of a BPSK spread communication system with BPSK or QPSK data modulation is (see Table 2)

$$\begin{aligned} P_{b,\,\text{barrage jamming}} &= Q\left(\sqrt{\frac{2E_b}{N_0 + T_c J/2}}\right) = Q\left(\sqrt{\frac{2PT_b}{N_0 + T_c J/2}}\right) \\ &= Q\left(\sqrt{\frac{2}{N_0/E_b + T_c J/(2PT_b)}}\right) \\ &= Q\left(\sqrt{\frac{2}{N_0/E_b + (J/P)(R/W)}}\right), \end{aligned} \tag{6.1}$$

where $W = 2/T_c$ is the null-to-null spread signal bandwidth (single-sided) and $R = 1/T_b$ is the bit rate.

Although the derivation is not quite as simple, it can be shown that basically the same expression holds if the jamming is partial band noise or single frequency [1]. In lieu of a detailed derivation, an approximate justification is that the de-spreader at the receiver front end, while dispreading the signal, spreads the partial band or single frequency jamming signal so that it appears as wideband Gaussian noise to the data demodulator. Similar arguments can be made for virtually any type of data modulation as long as the spreading is direct sequence, e.g., DPSK. Figure 23 illustrates BPSK/BPSK spread spectrum system performance in these types of jamming.

A somewhat more accurate analysis [13, 14], in the case of BPSK/BPSK, can be carried out for tone jamming of frequency equal to the carrier frequency and it shows that the interference component at the demodulator output is really binomially distributed, with the result that the bit error probability is

$$\begin{aligned} P_b &= Q\left[\sqrt{\frac{2}{N_0/E_b + (2J\,T_c/E_b \cos^2(\phi_J - \theta_s))}}\right] \\ &= Q\left[\sqrt{\frac{2}{N_0/E_b + (J/P)(R/W)\cos^2(\phi_J - \theta_s)}}\right], \end{aligned} \tag{6.2}$$

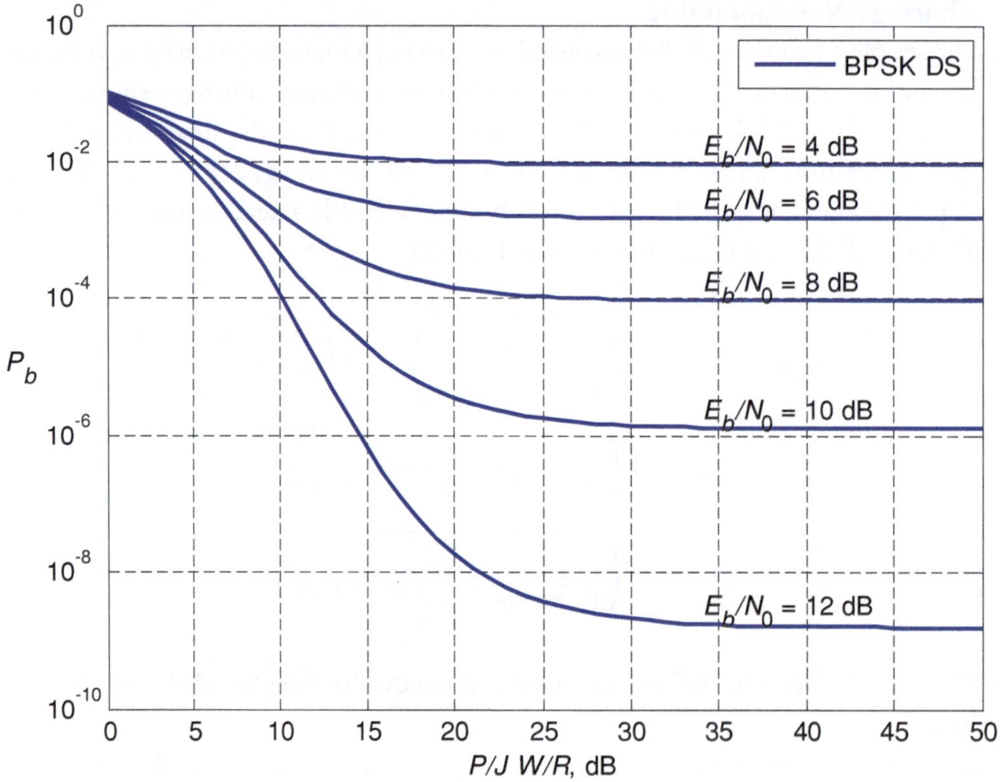

FIGURE 23: Performance of BPSK/BPSK spread spectrum in barrage, partial band, or tone jamming.

where $\phi_J - \theta_s$ is the phase difference between the jamming and signal. To get (6.2), the binomially distributed interference random variable was replaced by a Gaussian random variable with the same mean and variance. Note that if $\phi_J - \theta_s$ is an odd multiple of $\pi/2$, the term due to jamming is zero. If $\phi_J - \theta_s$ is an even multiple of $\pi/2$, (6.2) reduces to (6.1).

A similar analysis for QPSK spreading with BPSK data modulation can be carried out with the frequency offset of the jamming tone from the carrier frequency included. The result is

$$P_b = Q\left[\sqrt{\frac{2}{N_0/E_b + (J/P)(R/W)\,\mathrm{sinc}^2\,(\Delta f\, T_c)}}\right], \qquad (6.3)$$

where Δf is the frequency offset of the jamming from the signal. Note that if $\Delta f\, T_c$ is an integer, the jamming has no effect. Also note the lack of dependence on jammer phase relative to the signal phase.

As alluded to above, one could deduce the performance of FH spread spectrum in barrage noise jamming in a similar manner. For example, if the data modulation is noncoherent FSK, the expression for the bit error probability, from Table 2, is

$$P_b = \frac{M}{2(M-1)} \sum_{k=1}^{M-1} \frac{(-1)^{k+1}}{k+1} \binom{M-1}{k} \exp\left[\frac{-k \log_2 M}{k+1} \frac{E_b}{N_T}\right], \tag{6.4}$$

where, in the case of barrage jamming, $N_T = N_0 + N_J = N_0 + J/W$. Thus, E_b/N_T in (6.4) is replaced with

$$\frac{E_b}{N_T} = \frac{1}{N_0/E_b + (J/P)(R/W)}. \tag{6.5}$$

Results for $M = 2$ and 4 are given in Fig. 24.

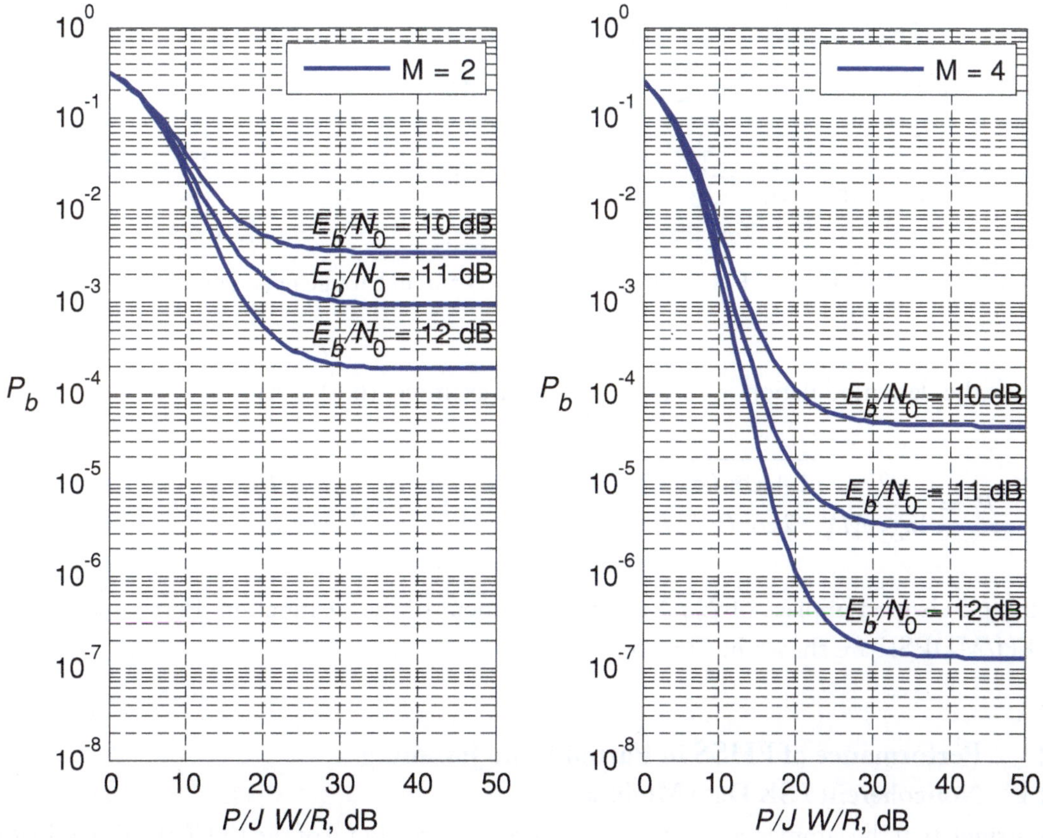

FIGURE 24: Performance of a FH/MFSK noncoherent spread-spectrum system in barrage noise jamming.

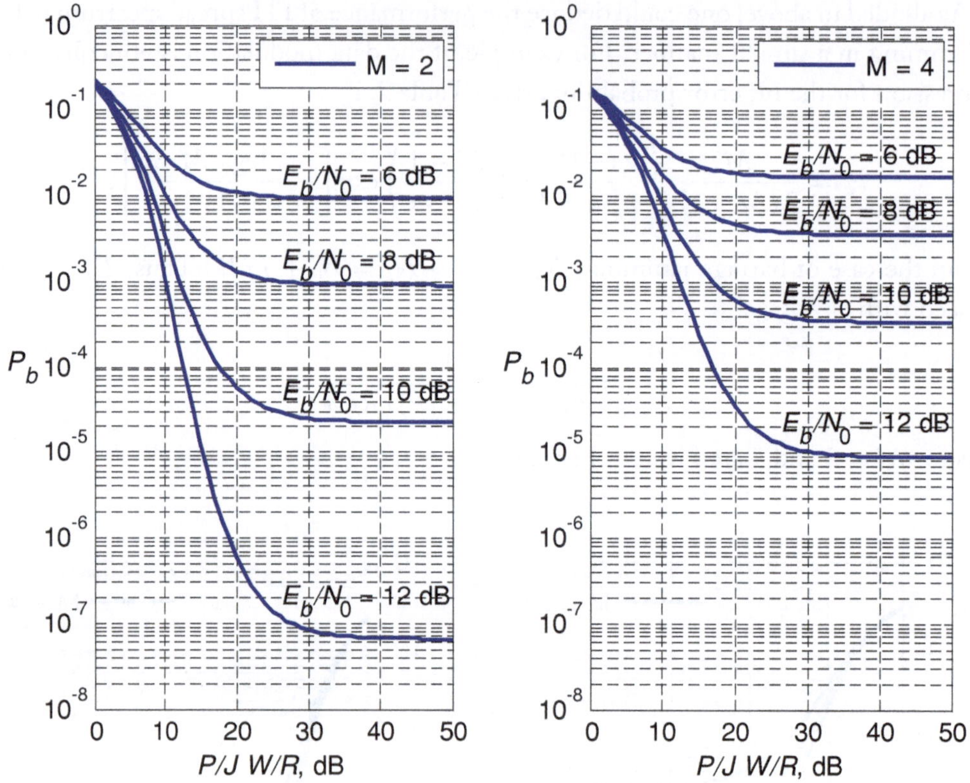

FIGURE 25: Performance of FH/MDPSK spread spectrum in barrage noise jamming.

If the data modulation is M-ary DPSK, for example, (6.4) is replaced by

$$P_b = \frac{2}{\log_2 M}\sqrt{\frac{1+\cos(\pi/M)}{2\cos(\pi/M)}}\,Q\left(2\log_2 M\left[1-\cos\left(\frac{\pi}{M}\right)\frac{E_b}{N_T}\right]\right)\quad M > 2, \tag{6.6}$$

where (6.5) is used in place of E_b/N_T in the argument of the Q-function. Performance curves for FH/MDPSK are shown in Fig. 25.

6.2 Performance of FHSS in Partial Band Jamming

6.2.1 Noncoherent FSK Data Modulation

We assume that the jammer concentrates its power in a fraction ρ of the FH/MFSK bandwidth. Thus, the jammer can disrupt data transmission whenever the transmitter hops into the jammed band while the jammer can concentrate its power in the jammed band. If ρ is the fraction of the spread bandwidth being jammed, the average probability of bit error is

$$\overline{P_b} = (1-\rho)\, P_b\left(\frac{P}{N_0 R}\right) + \rho\, P_b\left(\frac{1}{N_0/E_b + (J/P)(1/\rho)(R/W)}\right), \tag{6.7}$$

where

$$P_b = \frac{M}{2(M-1)} \sum_{k=1}^{M-1} \binom{M-1}{k} \frac{(-1)^{k+1}}{k+1} \exp\left(-\frac{k}{k+1}\frac{\log_2(M)\, E_b}{N_0}\right). \tag{6.8}$$

This can be differentiated with respect to ρ and set equal to zero to, in principle, solve for ρ_{opt}, which is the optimum (from the standpoint of the jammer) fraction of the spread bandwidth being jammed. Then ρ_{opt} can be substituted into (6.7) to produce a result for the worst-case bit error probability. The mathematics is somewhat complex since transcendental equations must be solved. Details are given in [2], [15] where it is shown that the worst-case bit error probability is

$$\left(\overline{P}_b\right)_{\max} = \frac{k'}{E_b/N_J}, \quad E_b/N_J \geq 2, \quad \begin{cases} k' = 0.3679 \text{ for } M = 2 \\ k' = 0.2329 \text{ for } M = 4 \\ k' = 0.1954 \text{ for } M = 8 \\ k' = 0.1812 \text{ for } M = 16. \end{cases} \tag{6.9}$$

Typical performance results are plotted in Fig. 26. Instead of the exponential decrease of bit error probability with E_b/N_J as in the case for Gaussian noise backgrounds, the optimum jammer imposes a decrease as the inverse of E_b/N_J. Note that the jammer must have knowledge of the communicator's signal-to-jamming energy ratio in order to impose this severe penalty.

6.2.2 DPSK Data Modulation

We again assume that the jammer concentrates its power in a fraction ρ of the FH/DPSK bandwidth. When not jammed, the noise power spectral density is $N_T = N_0$. When jammed, the noise power spectral density is $N_T = N_0 + N_J/\rho$. Thus, since $P_b = 0.5 \exp(-E_b/N_T)$ for binary DPSK, the average probability of error for the FH/DPSK system is

$$P_b = \frac{1}{2}(1-\rho)\exp(-E_b/N_0) + \frac{1}{2}\rho \exp\left[-\frac{1}{E_b/N_0 + (J/P)(R/W)(1/\rho)}\right]. \tag{6.10}$$

Following the previous procedure of finding the optimum ρ by differentiation of (6.10) and setting the result equal to zero, we find that

$$\rho_{\text{opt}} = \frac{1}{(P/J)(W/R)}. \tag{6.11}$$

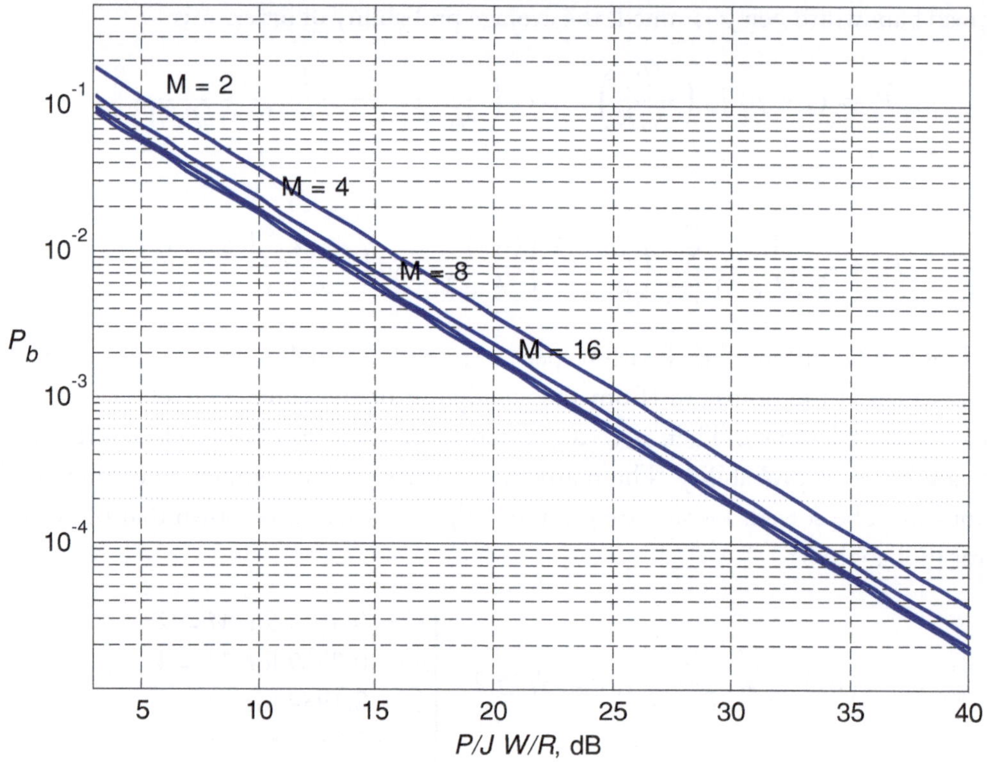

FIGURE 26: Performance of FH/MFSK noncoherent in worst-case partial band jamming.

When this value for ρ is substituted into (6.10), the worst-case (optimum from the jammer's standpoint) bit error probability is found to be

$$(\overline{P}_b)_{\max} = \begin{cases} \dfrac{e^{-1}}{2(P/J)(W/R)}, & \left(\dfrac{P}{J}\right)\left(\dfrac{W}{R}\right) \geq 1 \\ \dfrac{1}{2}\exp\left[-\left(\dfrac{P}{J}\right)\left(\dfrac{W}{R}\right)\right], & \left(\dfrac{P}{J}\right)\left(\dfrac{W}{R}\right) < 1. \end{cases} \tag{6.12}$$

This result is plotted in Fig. 27, where it is apparent that the partial band jamming imposes a severe penalty on system performance. Of course the jammer must know much about the system, in particular, the signal-to-jammer power ratio and the ratio of spread bandwidth to data rate.

6.3 Performance of DSSS with BPSK Data Modulation in Pulsed Jamming

The principle used by the jammer with FHSS can be used in DSSS. In the case of FHSS, the jammer concentrated its power over a fraction of the spread bandwidth. In the case of DSSS,

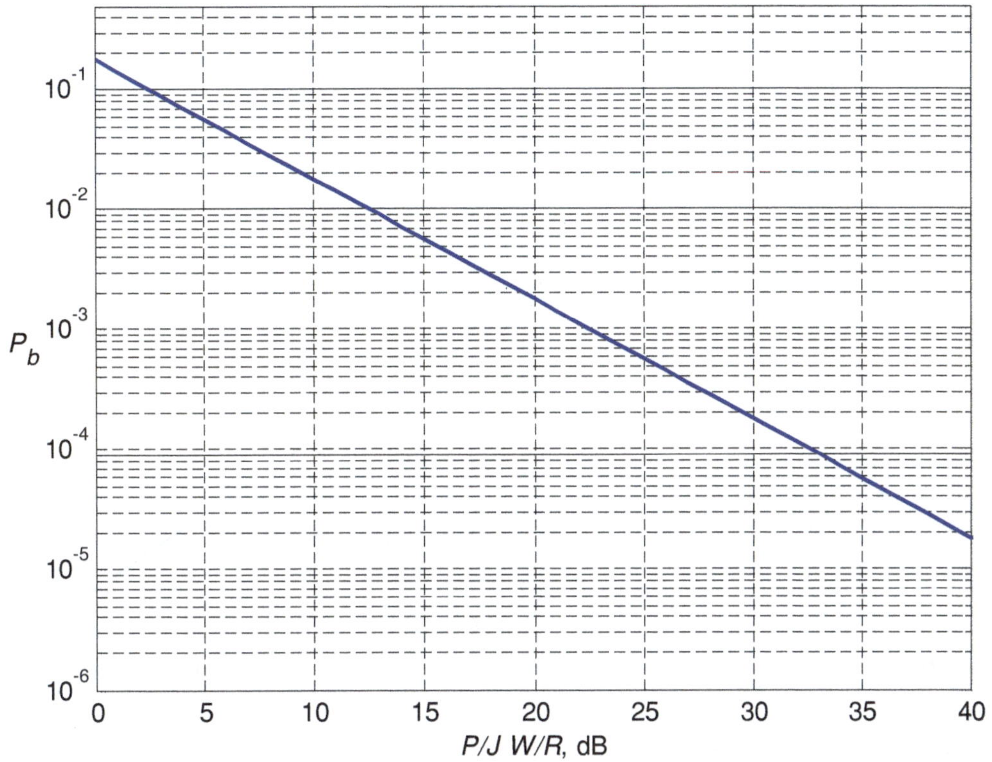

FIGURE 27: Bit error probability for the FH/DPSK spread spectrum in worst-case partial-band jamming.

the jammer can concentrate its power over a fraction of time. For BPSK data modulation, the average probability of bit error is

$$\overline{P}_b = \frac{1}{2}(1-\rho)\,Q\left(\sqrt{2E_b/N_0}\right) + \frac{1}{2}\rho\,Q\left[\sqrt{\frac{2}{E_b/N_0 + (J/P)(R/W)(1/\rho)}}\right]. \tag{6.13}$$

The usual procedure of differentiating with respect to ρ, setting the result equal to zero, solving for the optimum ρ, and back substituting into (6.13) can be followed to get the worst-case (from the communicator's standpoint) bit error probability. Using such numerical procedures, it can be shown that [2]

$$\left(\overline{P}_b\right)_{\max} = \begin{cases} \dfrac{0.083}{(P/J)(W/R)}, & \left(\dfrac{P}{J}\right)\left(\dfrac{W}{R}\right) \geq 0.709 \\ Q\left[\sqrt{2\left(\dfrac{P}{J}\right)\left(\dfrac{W}{R}\right)}\right], & \left(\dfrac{P}{J}\right)\left(\dfrac{W}{R}\right) < 0.709. \end{cases} \tag{6.14}$$

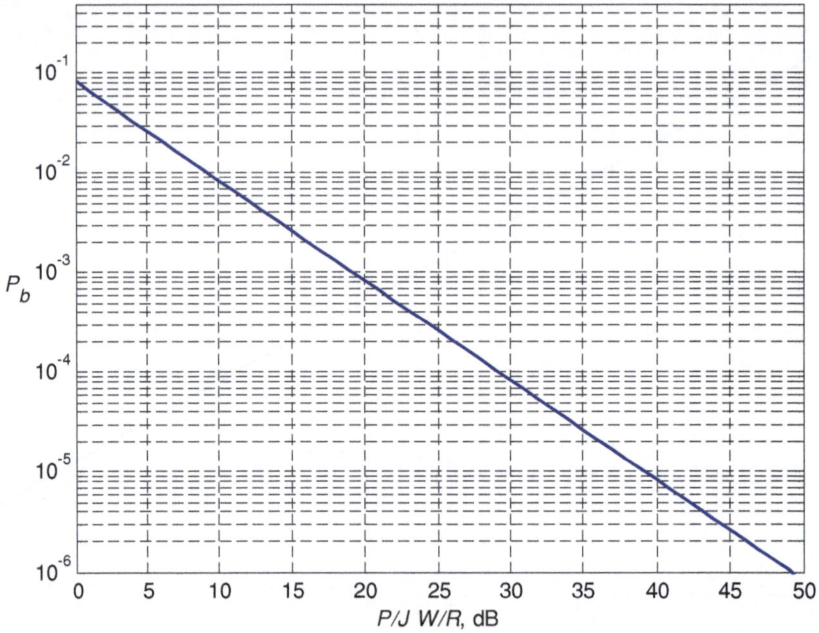

FIGURE 28: Worst-case bit error probability for DSSS with BPSK data modulation in pulse jamming.

Figure 28 shows the performance of DSSS/BPSK in pulse jamming where it is seen that the penalty due to worst-case jamming is similarly as severe as partial band jamming for FHSS. Again, as in the case of FHSS, the jammer must have considerable information about the communicator, in particular the signal-to-jamming power ratio and the bandwidth-to-bit-rate ratio, in order to impose this severe penalty.

6.4 Performance of FHSS in Multiple Tone Jamming

We now consider FHSS in multiple tone jamming. Two types of data modulation will be considered—MFSK and binary DPSK.

6.4.1 Noncoherent MFSK Data Modulation

The following assumptions hold in regard to noncoherent MFSK data modulation:

- The communication system is slow frequency hop; orthogonal frequency spacing.
- The jammer has complete knowledge of the receiver structure; a maximum of one jammer tone appears in each frequency hop slot.
- The MFSK tone spacing is $R_s = R/\log_2 M$, where R is the bit rate and, M is the number of FSK tones.
- The bandwidth of each hop slot is $W_d = MR/\log_2 M$.

- q tones are jammed, each with power $J_q = J/q$.
- Thermal noise is negligible.
- No symbol errors are made if $J_q < P$, where P is the average signal power.
- If $J_q > P$, an error is made if the tone frequency is within the FH band, but not in the same detector filter bandwidth as the signal.
- If $J_q = P$, a symbol error occurs with probability 1/2 under above conditions.
- The most FH bands are jammed if $J_q = P + \varepsilon$ and $q = \lfloor J/P \rfloor$ ($q_{\min} = 1$; $q_{\max} =$ integer part of W/W_d).
- The probability that any one frequency band is jammed is $P_J = q/(W/W_d)$.
- When a FH band is jammed and $q = \lfloor J/P \rfloor$, the symbol error probability is the probability that jammer and signal are not in the same band, or $\frac{M-1}{M}$. Hence,

$$P_s = \begin{cases} (M-1)/M, & W/W_d < \lfloor J/P \rfloor \\ [(M-1)/M]\, q W_d/W, & 1 \leq \lfloor J/P \rfloor \leq W/W_d \\ 0, & \lfloor J/P \rfloor < 1, \end{cases} \tag{6.15}$$

where $\lfloor \cdot \rfloor =$ the largest integer not exceeding the argument.

- The bit error probability is

$$P_b = \begin{cases} 0.5, & \left(\frac{P}{J}\right)\left(\frac{W}{R}\right) < \frac{M}{\log_2 M} \\ \frac{M}{2\log_2 M}\frac{1}{(P/J)(W/R)}, & \frac{M}{\log_2 M} \leq \left(\frac{P}{J}\right)\left(\frac{W}{R}\right) \leq \frac{W}{R} \\ 0, & \frac{W}{R} < \left(\frac{P}{J}\right)\left(\frac{W}{R}\right). \end{cases} \tag{6.16}$$

Bit error probability performance for FH/MFSK in multi-tone jamming is shown in Fig. 29. The reason for the steep decrease to $P_b = 0$ above a certain $(P/J)(W/R)$ is because of the assumption of no Gaussian noise and the finite amplitudes of the jamming tones.

6.4.2 Binary DPSK Data Modulation

The following assumptions hold in regard to DPSK data modulation:

- Assume binary DPSK modulation and slow frequency hop.
- Assume the following parameters:
  - $J =$ total jammer power
  - $q =$ number of jamming tones; $J_q = J/q$
  - $P =$ signal power

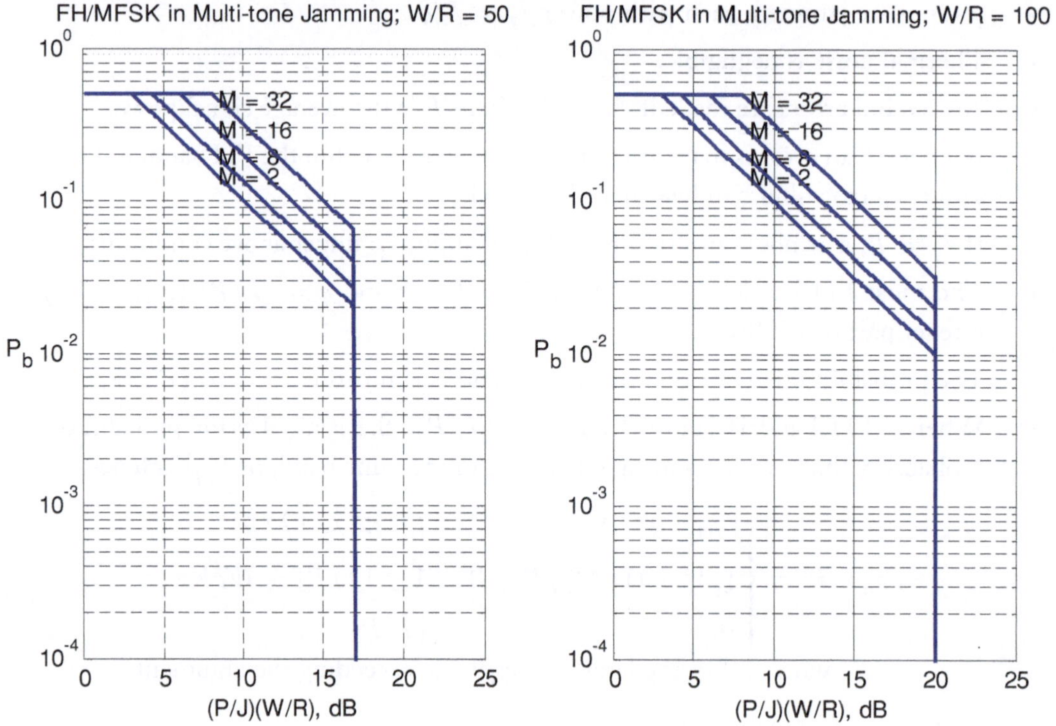

FIGURE 29: Performance of FH/MFSK in multitone jamming.

– $R =$ bit rate (the bandwidth of each FH band)
– $W =$ transmission bandwidth
– thermal noise is negligible

- $\rho = q/(W/R)$ is the probability of a particular band being jammed.
- Consider a single FH band being jammed. The DPSK demodulator compares phases of successive symbols. Let this phase difference be α. The possible decisions are:
  – $\pi/2 < \alpha \le \pi/2$, decide a 1 was transmitted
  – $\pi/2 < \alpha \le 3\pi/2$, decide a 0 was transmitted
- By considering phasor diagrams, we conclude that
  – for a 1 transmitted the receiver never makes an error,
  – for a 0 transmitted, $\cos\alpha = 2\left(J_q - P\right)/(R_1 R_2)$, where R_1 and R_2
  – are the signal plus jamming phasor sums during the present signaling
  – interval and the previous (reference) signaling interval; we conclude
  – that an error is made whenever $\cos\alpha \ge 0$ or when $J_q \ge P$;
  – a correct decision is made if $J_q < P$.

- The optimum jamming strategy is to put just enough power on each tone, so an error is made when a 0 is transmitted for a jammed frequency hop:

  – The optimum number of tones is $q = \lfloor J/P \rfloor$.
  – The average bit error probability is
  $P_b = 1/2 \, \Pr(\text{error} \mid 0 \text{ trans}) + 1/2 \, \text{pr}(\text{error} \mid 1 \text{ trans}) = 1/2 \, \Pr(\text{error} \mid 0 \text{ trans})$.

- The conditional P_b given that a 0 was transmitted is ρ (the probability that a hop frequency is jammed).
- Putting this all together, we have

$$P_b = \frac{1}{2}\frac{q}{W/R} \approx \frac{1}{2}\frac{J/R}{W/R} = \frac{1}{2(P/J)(W/R)}, \; 1 \le q \le W/R. \tag{6.17}$$

- For large jammer power, all FH bands can be jammed and $P_b = 1/2$.
- When $J < P$, not even a single band can be jammed so $P_b = 0$.

Thus, collecting these results, we have

$$P_b = \begin{cases} \frac{1}{2}, & \left(\frac{P}{J}\right)\left(\frac{W}{R}\right) < 1 \\ \frac{0.5}{(P/J)(W/R)}, & 1 \le \left(\frac{P}{J}\right)\left(\frac{W}{R}\right) < \frac{W}{R} \\ 0, & \frac{W}{R} \le \left(\frac{P}{J}\right)\left(\frac{W}{R}\right). \end{cases} \tag{6.18}$$

Performance results for FH/DPSK in multi-tone jamming are shown in Fig. 30. The reasons for the abrupt drop of the bit error probability to zero above a certain $(P/J)(W/R)$, as in the case of FH/MFSK, is because of the assumption of no Gaussian noise and the finite amplitudes of the jamming tones.

6.5 Conclusions

The cases analyzed above demonstrate that severe performance penalties can be imposed on spread spectrum systems in the face of jamming. These penalties are very severe for partial band or pulsed jammers that have sufficient knowledge of the received signal, namely, the signal-to-jamming power ratio and the spread signal bandwidth relative to the data rate. In the next section, it will be shown that this performance degradation can be combated, to a large degree, by the application of forward error correction coding.

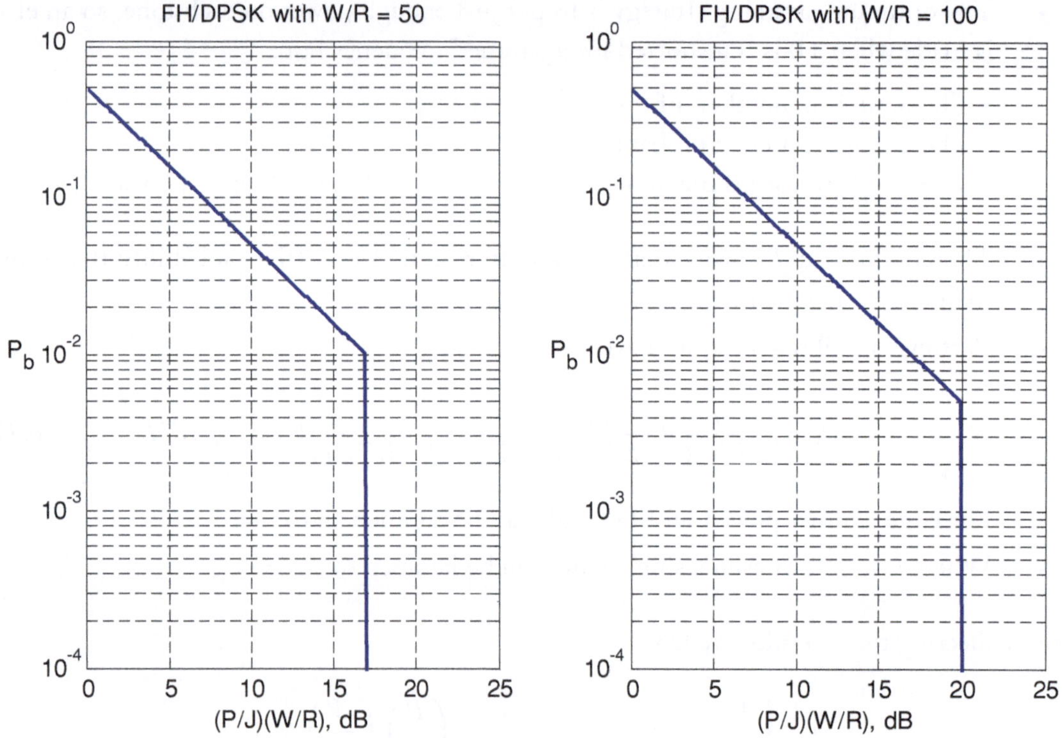

FIGURE 30: Performance of FH/DPSK in multitone jamming.

7 PERFORMANCE OF SPREAD SPECTRUM SYSTEMS OPERATING IN JAMMING WITH FORWARD ERROR CORRECTION CODING

Forward error correction coding introduces redundancy into a stream of information symbols (bits) by the inclusion of check symbols. This redundancy hopefully improves the overall error probability even though less energy per encoded symbol is used because the encoded symbol rate must be higher than that of the unencoded symbol stream in order to maintain the same information rate with coding as without.

We can identify the following mechanisms for the improvement of the performance of spread spectrum systems in jamming environments:

- Spreading of the interference at the receiver front end by multiplication by the local de-spreading code.
- Use of forward error correction coding (FEC) in conjunction with interleaving/deinterleaving.

- Partial knowledge of the jammer (jammer state information).

It is the purpose of this section to summarize some results regarding the performance improvement due to coding of spread spectrum systems operating in jamming environments.

As far as coding is concerned, the waveform channel can be replaced by a discrete memoryless channel (DMC). Even if the channel introduces memory, it is assumed that interleaving at the transmitter and deinterleaving at the receiver is used to essentially remove the channel-induced memory. A DMC is characterized by the *a posteriori* probability $p(\mathbf{y} \mid \mathbf{x}_m, \mathbf{z})$, where $\mathbf{y}$ is the channel output, $\mathbf{x}_m$ is the mth codeword, and $\mathbf{z}$ is a RV describing the jammer state. The impact of jammer state information will not be considered in this summary. We will summarize results for two different types of coding: block and convolutional.

7.1 Block Coding Concepts

When an (n, k) block code is referred to, this means that a block of k information symbols (bits) is encoded into a block of $n > k$ encoded symbols. The added $n - k$ symbols are called *parity symbols* and the *code rate* is $R = k/n$. For the mth codeword,

– the vector $\mathbf{w}_m = w_{m0}\ w_{m1} \cdots w_{m(k-1)}$ represents the information sequence;
– the vector $\mathbf{x}_m = x_{m0}\ x_{m1}\ \cdots\ x_{m(n-1)}$ represents the codeword;
– the encoder performs a one-to-one mapping of $\mathbf{w}_m$ into $\mathbf{x}_m$.

Observe that for the case of binary information and encoded sequences, that a fraction $2^k/2^n = 2^{k-n} = 2^{k(1-1/R)}$ of all possible codewords are used, which decreases as k increases for $R < 1$. Therefore, the codewords composing the code can be "more spread out" the larger k if the codewords are selected with care.

Two useful concepts are Hamming distance and Hamming weight.

- The Hamming distance between two codewords is the number of 1s in the modulo-2 sum of the codewords. For example, given 1101000 and 1011100, for which the modulo-2 sum is 0110100, results in a Hamming distance of 3.
- The Hamming weight of a codeword is defined as the number of 1s in a codeword. For example, the Hamming weight of 1101000 is 3.

If the minimum Hamming distance between all codewords in a code is $d_{\min}$, then up to $\lfloor d_{\min} - 1 \rfloor /2$ errors can be corrected, where $\lfloor \cdot \rfloor$ denotes the largest integer not exceeding the argument. For example, a code with minimum distance 3 can correct one error.

Some terms applying to block codes are

- A *systematic* code is one where the message vector appears directly in the code vector.

- For a *linear* code the sum of any two codewords is also a codeword. This means that we can characterize the error-correction capability of a linear code by its *weight distribution*; this is the set A_d of all codewords a distance d from the all-0s codeword. For example, a code with weight distribution $A_3 = 7$, $A_4 = 7$, and $A_7 = 1$ has seven codewords distance 3, seven codewords distance 4, and one codeword distance 7 from the all-0s codeword.
- A *cyclic* code is one where the codeword are cyclic shifts of each other.

It can be shown that the optimum decoding rule is the following:

- For hard decisions (i.e., 1-0 decisions), choose as the transmitted codeword the one closest to the received data vector with distance measured in the Hamming sense.
- For soft decision information, choose as the transmitted codeword the closest to the received data vector with distance measured in the Euclidian sense; i.e., choose $\mathbf{x}_m$ to minimize

$$\sum_{j=1}^{n} \left(y_j - x_{m_j}\right)^2, \tag{7.1}$$

where $\mathbf{y} = [y_1, y_2, \ldots, y_m]$ is the received data vector.

Bit error probabilities for general linear codes are difficult to compute exactly. For bounded distance decoders (i.e., decoders that can correct up to E errors and no more), it has been argued [14] that the bit error probability is bounded by

$$P_b = \frac{1}{k} \sum_{i=E+1}^{n} \min\left[k, i + E\right] \binom{n}{i} p^i (1-p)^{n-i}, \tag{7.2}$$

where p is the probability of a symbol error in the received codeword. For the jamming channels of the previous section, p would be one of the bit error probability expressions given there.

7.1.1 BCH Block Codes [16]

An example family of block codes is the Bose—Chaudhuri–Hocquenghem (BCH) codes, which are linear cyclic codes. Nonbinary BCH codes exist, but we will limit our attention to only the binary. The block length for binary BCH codes is always $n = 2^m - 1$, $m \geq 3$ an integer. The number of errors that can be corrected is bounded by $t < (2^m - 1)/2$ and it is always true that $n - k \leq mt$. Specific values for t and k are given in Table 11. Approximate bit error probabilities versus E_b/N_0 in Gaussian noise channels can be computed using (7.2) by using the appropriate bit error probability expression from Table 2. In doing so, it is important to remember to replace E_b/N_0 in these expressions with RE_b/N_0, where $R = k/n$ to account for the increase in code symbol rate required to keep the overall bit rate the same (n code symbols are required for each k bits sent through the channel). To determine coded bit error rate in spread spectrum with jamming, the same procedure is used, but the appropriate bit error probability expression derived in Section 5.4 is used instead. This will be done later after other coding techniques are discussed.

7.1.2 Reed–Solomon Block Codes [16]

Another important family of block codes are Reed–Solomon codes. They are nonbinary block codes that have found important applications in space communications and compact disc technology, among other applications. Reed–Solomon codes are particularly effective in applications where errors tend to occur in bursts.

Reed–Solomon codes use alphabets having 2^m symbols, $\{0, 1, \ldots, 2^m - 1\}$, with block length $n = 2^m - 1$. The codes can correct up to e_0 errors with the number of parity symbols being $n - k = n - 2e_0 = 2^m - 1 - 2e_0$. The minimum distance of this code family is $d_{\text{min}} = 2e_0 + 1$, where the Hamming distance between nonbinary codewords is defined to be the number of positions in which the codewords differ. Reed–Solomon codes are often used in channels that are nonbinary, for example, ones where M-ary FSK is the modulation scheme of choice (if 8-FSK is used, it would be convenient to choose an $m = 3$ Reed–Solomon code). If used in a binary channel, bits may be grouped to form m-bit blocks. In this case, the Reed–Solomon code can be thought of as accepting $k' = km$ information bits and mapping them into channel symbol blocks of length $n' = nm$ binary channel symbols. Thus, the rate of the Reed–Solomon code is

$$R = \frac{k'}{n'} = \frac{k}{n} = \frac{2^m - 1 - 2e_0}{2^m - 1} \tag{7.3}$$

or

$$e_0 = \left\lceil (1 - R)\frac{2^m - 1}{2} \right\rceil. \tag{7.4}$$

The probability of bit error for Reed–Solomon codes is over bounded by

$$P_b \leq \sum_{i=e_0+1}^{2^m-1} \frac{i}{2\left(2^m - 2\right)} \binom{2^m - 1}{i} p_s^i \left(1 - p_s\right)^{2^m-1-i}, \tag{7.5}$$

where, for noncoherent MFSK, the symbol error probability is given by

$$p_s = \sum_{k=1}^{M-1} \frac{(-1)^{k+1}}{k+1} \binom{M-1}{k} \exp\left[\frac{-k}{k+1}\frac{E_s}{N_0}\right] \tag{7.6}$$

with the symbol energy-to-noise spectral density ratio for Gaussian noise being given by $E_s/N_0 = mR(E_b/N_0)$ which accounts for the code rate $R = k/n$ and the fact that m binary bits are associated with one 2^m-ary symbol. If a binary modulation scheme is used, (7.6) would be replaced by

$$p_s = 1 - (1 - p)^m, \tag{7.7}$$

where p is the probability of bit error for the appropriate binary modulation scheme. For example, for BPSK in white Gaussian noise it is

$$p = Q\left(\sqrt{\frac{2RE_b}{N_0}}\right). \tag{7.8}$$

Results for spread spectrum communication system performance in jamming using Reed–Solomon codes will be presented later.

7.2 Convolutional Codes

Convolutional codes differ from block codes in that the information bits are not grouped into blocks for encoding, but rather a linear shift-register circuit is used to map a continuous sequence of input symbols (bits) into a continuous sequence of output symbols (bits). The principle of keeping the allowed codewords separated in Hamming distance as much as possible still holds as it does for block codes. A convolutional code can be characterized in various ways, including an encoder block diagram, its code generators, a state transition diagram, or a trellis diagram. Figure 31 shows an example block diagram of a convolutional encoder, where the adders are modulo-2 and the input information bits are clocked in at the left in time sequence. For each input bit, two output bits are generated because the switch on the right-hand side first is in the upper position and then flips to the lower position for each input bit. For the input $\{101\} \Rightarrow 1 + D^2$, for example, we have at the upper adder output

$$(1 + D^2)(1 + D + D^2) = 1 + D + D^3 + D^4 \Rightarrow \{11011\}$$

and we have at the lower adder output

$$(1 + D^2)(1 + D^2) = 1 + D^4 \Rightarrow \{10001\},$$

where the arithmetic is modulo-2. Sampling a bit from the upper leg and then from the lower gives output encoded sequence {1110001011}. We will not exhibit the state transition diagram or the trellis diagram for this encoder. For a rate-1/3 code, there would be three adders in the block diagram and the output would sample sequentially from the outputs of these three adders. From the encoder block diagram, it should be clear that convolutional codes are linear.

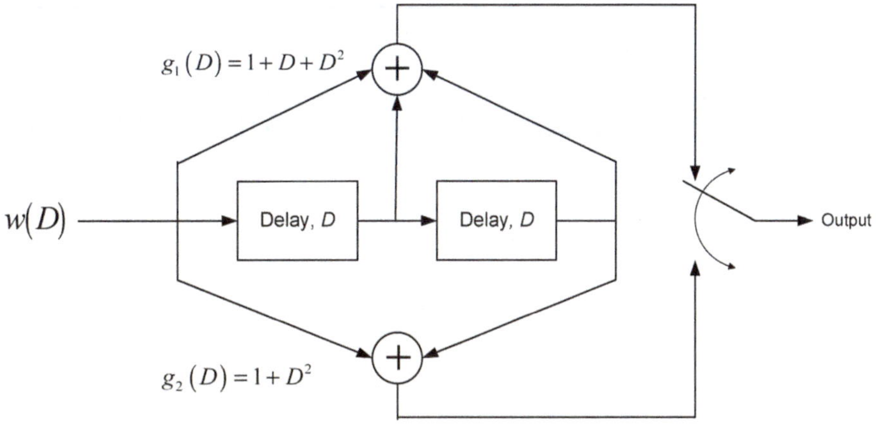

FIGURE 31: Block diagram of a rate-1/2 convolutional encoder.

Maximum likelihood decoding of a convolutionally encoded sequence in noise is performed by the Viterbi algorithm. Performance of a convolutional code is analyzed by finding the probability of deviating from the correct path through the trellis and determining the resulting number of bit errors. The probability of bit error for a convolutional code is over bounded by

$$P_b < \sum_{k=d_{\text{free}}}^{\infty} c_k P_k, \tag{7.9}$$

where d_{free}, called the free distance, is the Hamming distance between the all-zeros path in the trellis and the minimum-length path deviating from it, and

$$\begin{aligned} P_k &= \sum_{e=k/2+1}^{k} \binom{k}{e} p^e (1-p)^{k-e} + \frac{1}{2} p^{k/2} (-p)^{k/2}, \; k \text{ even} \\ P_k &= \sum_{e=(k+1)/2}^{k} \binom{k}{e} p^e (1-p)^{k-e}, \; k \text{ odd} \\ p &= \text{hard decision channel error probability} \end{aligned} \tag{7.10}$$

for hard (1-0) channel decisions, and

$$P_k = Q\left(\sqrt{\frac{2kRE_b}{N_0}}\right) \tag{7.11}$$

for soft channel decisions assuming BPSK signaling in additive Gaussian noise backgrounds. The constants c_k can be found by computer simulation for a given convolutional code and are listed in Tables 12 and 13 for the best rate-1/2 and rate-1/3 codes, respectively.

TABLE 11: Abbreviated List of BCH Code Parameters [16]

n	k	t	n	k	t	n	k	t
31	21	2	255	247	1	511	502	1
	16	3		231	3		484	3
	11	5		223	4		466	5
63	57	1		215	5		448	7
*	45	3		207	6		430	9
&	30	6	*	191	8		412	11
	24	7		163	12	*	385	14
%	16	11		147	14		358	18
127	120	1	&	131	18		322	22
	106	3		115	21	&	259	30
*	99	4		99	23		211	41
	78	7		87	26		175	46
&	64	10	%	63	30	%	130	55
	50	13		55	31		103	61
	36	15		45	43		67	87
%	29	21		29	47		31	109
	8	31		13	59		10	121

$* \simeq$ rate 3/4.
& $\simeq$ rate 1/2.
% $\simeq$ rate 1/4.

7.3 Example System Performances for Spread Spectrum Systems with Coding Operating in Jamming Environments

In this section, three example systems are considered to show the improvement afforded by coding in jammed spread spectrum systems.

TABLE 12: Best Rate-1/2 Convolutional Codes and Their Partial Weight Structure [16]

CONSTR. LENGTH, v	CODE GENER-ATORS (OCTAL)	FREE DIST-, ANCE d_f	c_k FOR $d =$ d_f	d_f+1	d_f+2	d_f+3	d_f+4	d_f+5	d_f+6	d_f+7
3	(7, 5)	5	1	4	12	32	80	192	448	1024
4	(15, 15)	6	2	7	18	49	130	333	836	2069
5	(35, 23)	7	4	12	20	72	225	500	1,324	3680
6	(75, 53)	8	2	36	32	62	332	701	2,342	5503
7	(171, 133)	10	36	0	211	0	1404	0	11,633	0
8	(371, 247)	10	2	22	60	148	340	1008	2,642	6748
9	(753, 561)	12	33	0	281	0	2179	0	15,035	0

TABLE 13: Best Rate-1/3 Convolutional Codes and Their Partial Weight Structure [16]

CONSTR. LENGTH, v	CODE GENER-ATORS (OCTAL)	FREE DIS-TANCE, d_f	c_k FOR $d =$ d_f	d_f+1	d_f+2	d_f+3	d_f+4	d_f+5	d_f+6	d_f+7
3	(7, 7, 5)	8	3	0	5	0	58	0	201	0
4	(17, 15, 13)	10	6	0	6	0	58	0	118	0
5	(37, 33, 25)	12	12	0	12	0	56	0	320	0
6	(75, 53, 47)	13	1	8	26	20	19	62	86	204
7	(171, 145, 133)	14	1	0	20	0	53	0	184	0
8	(367, 331, 225)	16	1	0	24	0	113	0	287	0

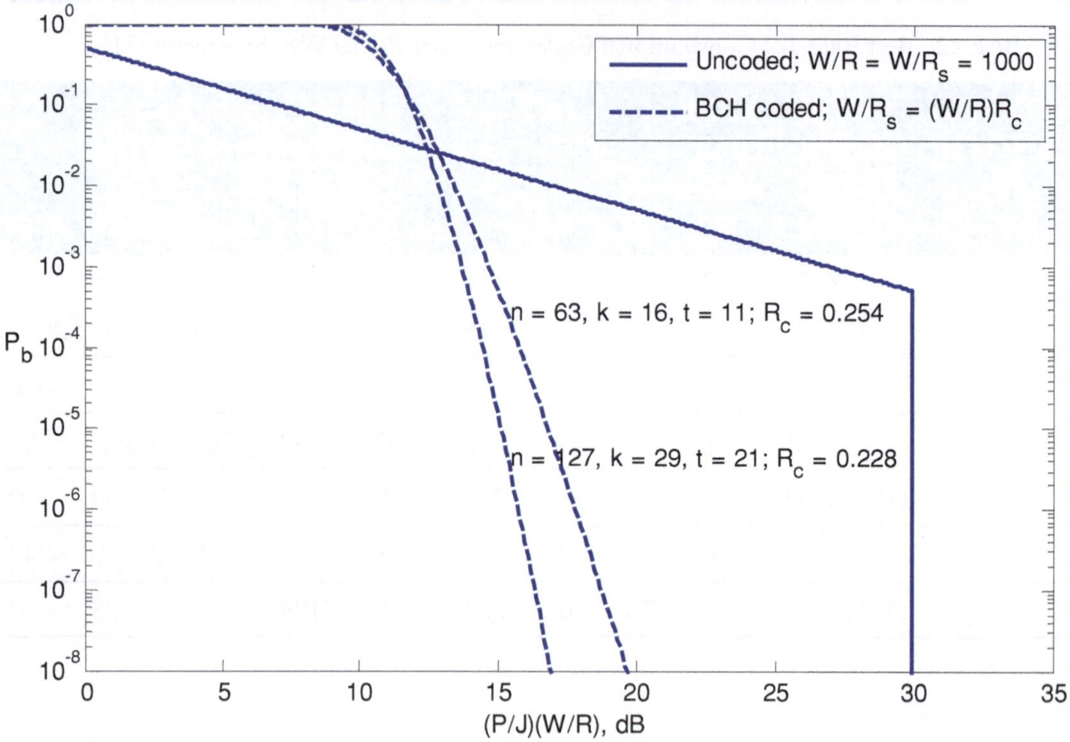

FIGURE 32: Rate-1/4 BCH codes to improve the performance of FH/DPSK in optimum tone jamming.

Example 11. Consider a FH/DPSK spread spectrum system in worst-case tone jamming with $W/R = 1000$. Investigate the use of rate-1/4 and rate-1/2 BCH coding to improve performance.

Solution: The channel symbol error probability is given by (6.18). The bit rate and symbol rate are related by $R_s = nR/k$. A MATLAB program for computing performance, based on (7.2), was used to obtain the results shown in Fig. 32 for codes of approximately rate 1/4 and in Fig. 33 for codes of approximately rate 1/2. Note that the rate-1/4 codes actually outperform the rate-1/2 codes due to their greater error-correction capability. This is in spite of the fact that W/R_s is less for the rate-1/4 codes than for the rate-1/2 codes (i.e., the former case provides less protection due to spreading than the latter).

Example 12. Consider a FH/DPSK spread spectrum system in worst-case tone jamming with $W/R = 1000$. Investigate the use of Reed–Solomon coding to improve performance.

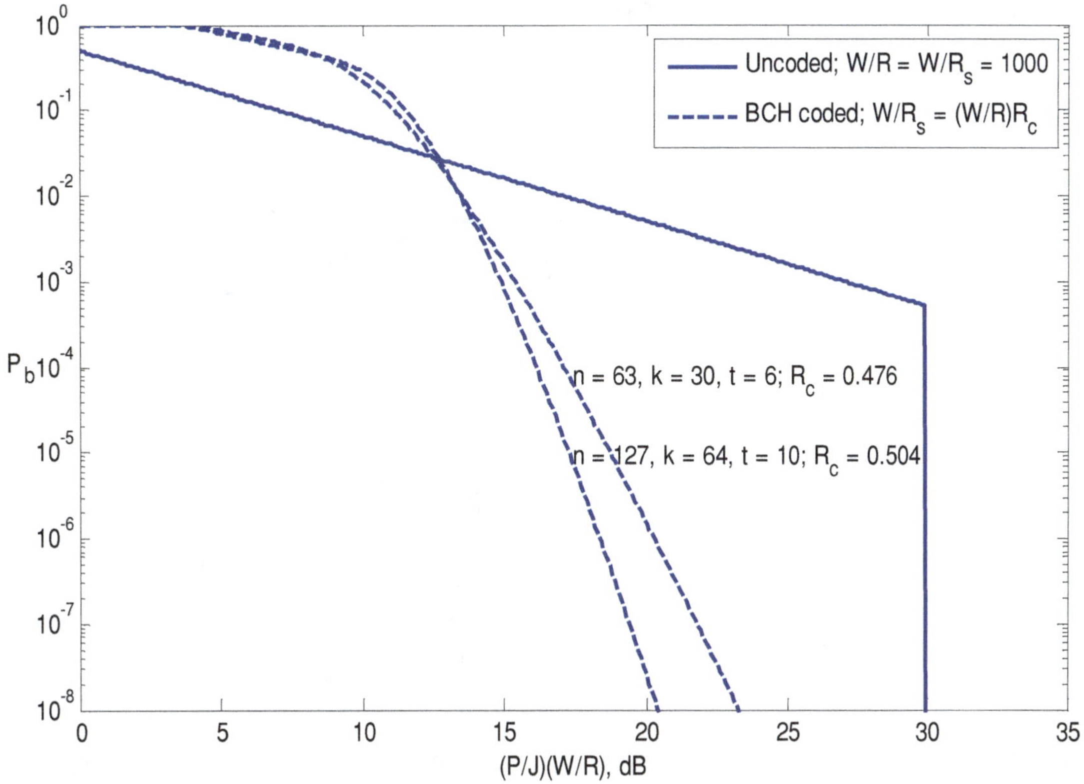

FIGURE 33: Rate-1/2 BCH codes to improve the performance of FH/DPSK in optimum tone jamming.

Solution: The channel symbol error probability is given by (6.18). A MATLAB program for computing performance, based on (7.5), was used to produce the plots given in Fig. 34. Again note that the rate-1/4 code slightly outperforms the rate-1/2 code [at high $(P/J)(W/R)$] due to its greater error-correction capability. This is in spite of the fact that W/R_s is less for the rate-1/4 code than for the rate-1/2 code.

Example 13. Consider a FH/DPSK spread spectrum system in worst-case tone jamming with $W/R = 1000$. Investigate the use of convolutional coding to improve performance.

Solution: The channel symbol error probability is given by (6.18). A MATLAB program for computing performance, based on the bound of (7.9), was used to produce the results given in Fig. 35 for the rate-1/2 code and in Fig. 36 for the rate-1/3 code. Since the modulation is binary DPSK, bits were blocked into 6-bit blocks with the symbol error probability computed from (7.7). The code symbol energy is related to the bit energy by $E_s = kE_b/n$.

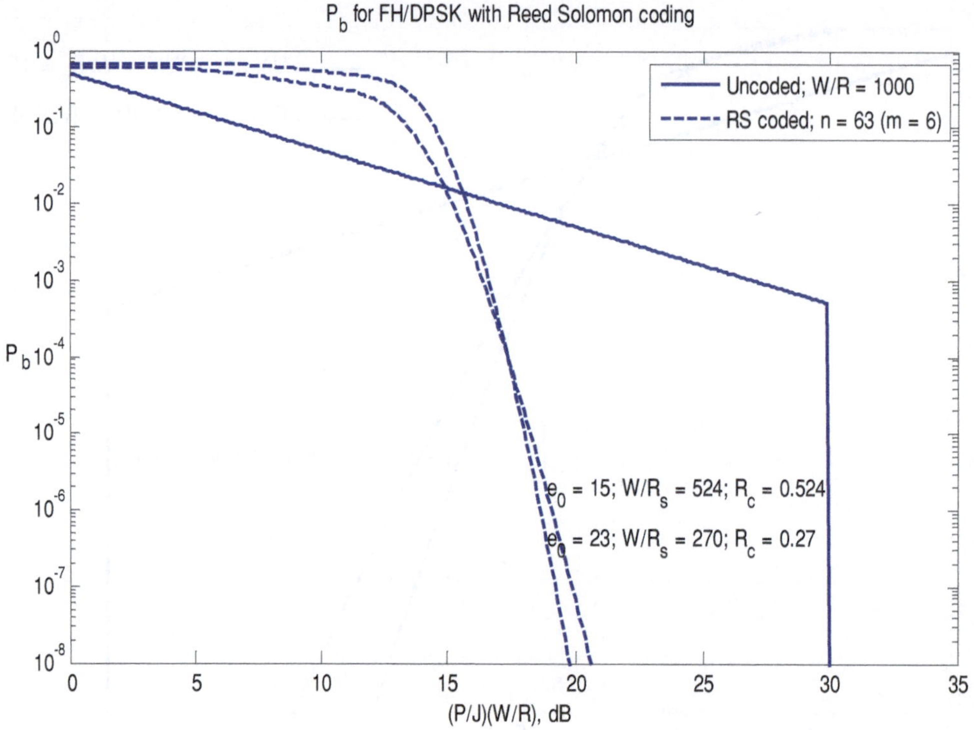

FIGURE 34: Use of Reed-Solomon coding to improve the performance in FH/DPSK with optimum tone jamming.

7.4 Summary

Although the penalties imposed by optimized jammers can be very severe, the above examples have shown that both block and convolutional coding can be used to combat much of the performance degradation imposed by jamming. It is emphasized that an implicit assumption in the use of such codes, which work most effectively if the errors are randomly distributed, is that any tendency for the errors to be bunched is combated by use of appropriately designed interleaving at the transmitter and corresponding de-interleaving at the receiver. The results given in Figs. 32–36 indicate that improvements on the order of 15–20 dB can be expected at bit error probabilities of 10^{-3} or lower.

8 PERFORMANCE IN MULTIPLE USER ENVIRONMENTS

As discussed in conjunction with Fig. 1, more than one user occupying the same time–frequency space can exist simultaneously in a spread spectrum system if their respective spreading codes have low cross-correlation between them. This property of spread spectrum systems is called code-division multiple-access (CDMA) capability. To investigate some of the aspects of this

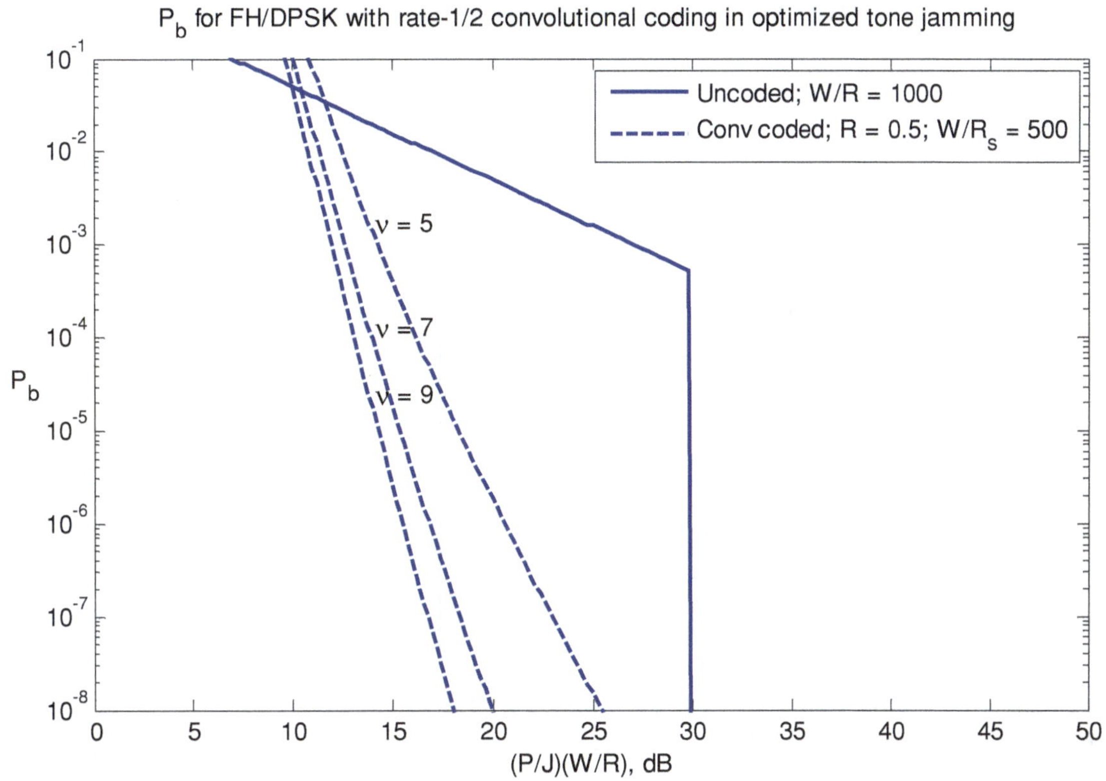

FIGURE 35: Rate-1/2 convolutional coding to improve performance of FH/DPSK in optimized tone jamming.

capability of spread spectrum systems, consider the simplified block diagram of Fig. 37, which represents a baseband version of K simultaneous users transmitting data streams by using spreading codes, presumably well chosen so that their mutual correlations are low. The different delays account for possible differences in propagation times for different users. AWGN is introduced primarily by the receiver front end of the intended receiver (in this case that of user 1 who correlates with its code and integrates over the bit interval). The received signal for user 1 is written as

$$y(t) = A_1 d_1 (t - \tau_1) c_1 (t - \tau_1) + \sum_{k=2}^{K} A_k d_k (t - \tau_k) c_k (t - \tau_k) + n(t), \tag{8.1}$$

where A_k and τ_k, $k = 1, 2, \ldots, K$, represent the amplitude and delay, respectively, for the kth user. Assuming perfect synchronization of the local code for user 1, we can take $\tau_1 = 0$ and

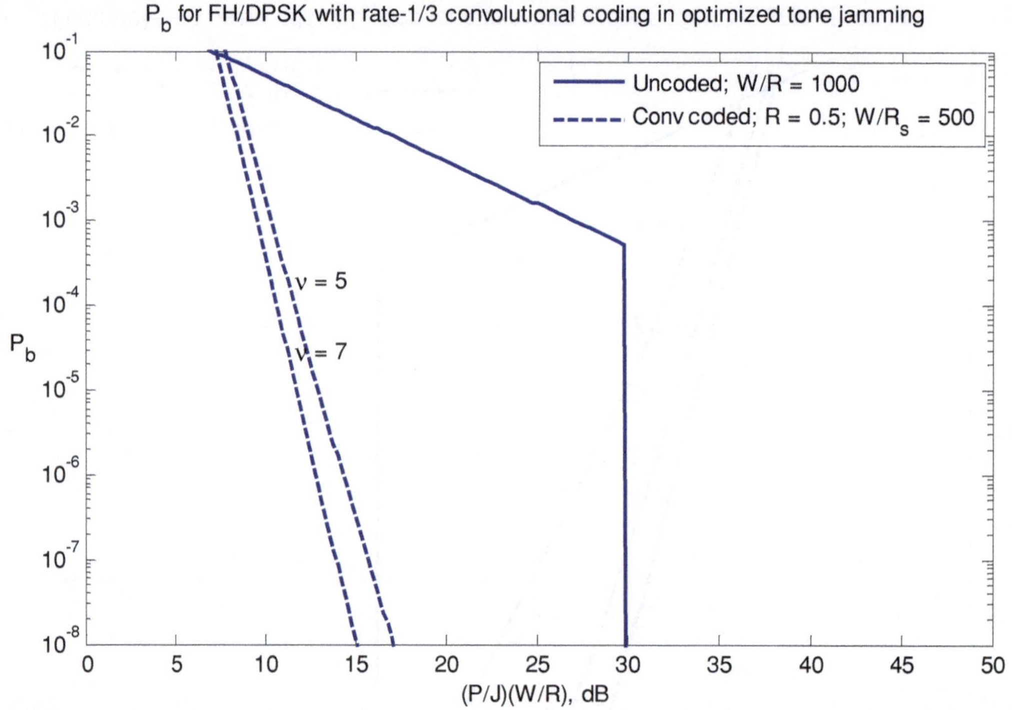

FIGURE 36: Rate-1/3 convolutional coding to improve the performance of FH/DPSK in optimized tone jamming.

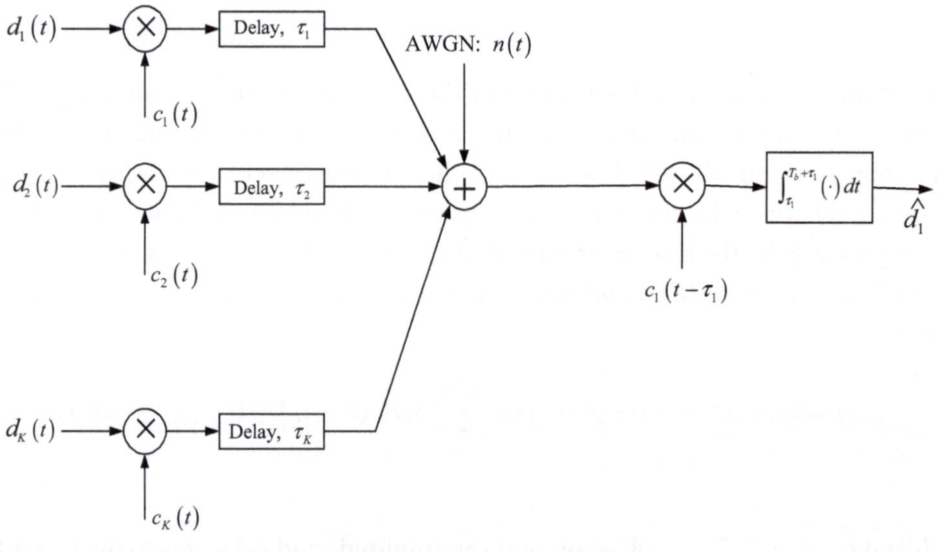

FIGURE 37: Block diagram of a CDMA system.

thereby write the output of the integrator (representing the detector for user 1) as

$$Y = A_1 d_1(0)\, T_b + \sum_{k=2}^{K} A_k T_b d_k(0)\, \rho_{1k} + N_g, \tag{8.2}$$

where

$$\rho_{1k} = \frac{d_k(-1)}{d_k(0)} \int_0^{\tau_k} c_1(t)\, c_k(t + T_b - \tau_k)\, \mathrm{d}t + \int_{\tau_k}^{T_b} c_1(t)\, c_k(t - \tau_k)\, \mathrm{d}t; \qquad |\tau_k| \le T_b.$$

In (8.2), the first term is the desired correlator output due to user 1, the second sum of terms is referred to as multiple access noise with ρ_{1k} being the aperiodic correlation of the receptions from user 1 and user k, and N_g is a Gaussian random variable due to integration of the AWGN. Note that if an orthogonal code set were used and it were possible to maintain perfect synchronism between codes through the channel, the multiple access noise would be zero. This is almost never the case, however, even if an orthogonal code set is used, due to timing misalignments and multipath.

Several approaches to calculating the performance of CDMA reception have been published over the years. The simplest of these rely on the Central Limit Theorem to approximate the multiple access noise terms as Gaussian [17], which almost always result in optimistic performance results. An extensive study on such approximations is given in [18]. For our purposes here, we quote such a result from [19]

$$P_{b,\mathrm{MAI}} = Q\left[\sqrt{\left(\frac{N_0}{2T_b P_1} + \frac{1}{3N}\sum_{j=2}^{K}\frac{P_j}{P_1}\right)^{-1}}\,\right] = Q\left[\sqrt{\left(\frac{N_0}{2E_{b1}} + \frac{1}{3N}\sum_{j=2}^{K}\frac{P_j}{P_1}\right)^{-1}}\,\right], \tag{8.3}$$

where

E_{b1} = bit energy of user of interest (user 1),
T_b = bit duration,
P_j = average power of user j,
N_0 = power spectral density of the AWGN,
N = number of code chips per bit,
K = total number of active users.

Figure 38 shows the performance of a 10-user system, with all users assumed to be 1 W except user 2 whose power varies as shown on the graph, and a processing gain of 127. Two system characteristics may be noted: (1) the presence of other users means that a floor, which is due to the multiple-access interference, for the bit error probability is eventually approached;

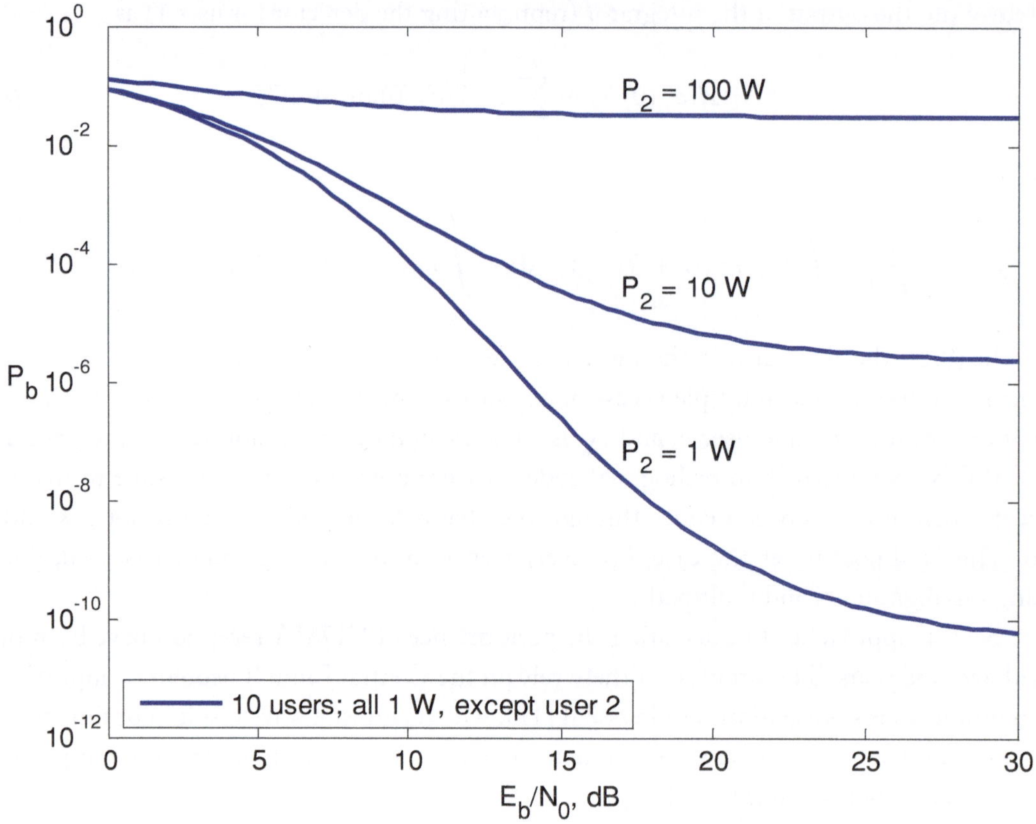

FIGURE 38: Multiple-access performance for 10 user system; processing gain of 127.

(2) user 2 powerful enough means that it eventually dominates the system performance. The latter phenomenon is referred to as receiver capture, in this case by user 2, and is characteristic of CDMA communication systems. For this reason, power control is invariably used in CDMA systems to guard against domination of the system performance by a single user.

9 SUMMARY

In this lecture, the fundamental concepts of spread spectrum modulation have been presented. Spread spectrum modulation can be defined as any modulation scheme that utilizes a transmission bandwidth much greater than the modulating signal bandwidth, *independently* of the bandwidth of the modulating signal. After describing the generic forms of spread spectrum, known as direct sequence and frequency hop, the subject of spreading code generation was surveyed. Codes considered were m-sequences, Gold codes, quaternary sequences, Kasami sequences, and Walsh sequences. Properties of m-sequences were illustrated by example. Next, the important topic of code acquisition at the receiver was discussed with two types of acquisition described—serial search and matched filter. Two circuits for code tracking were briefly described next—the delay-lock and tau-dither tracking loops. The former exhibits a slightly smaller tracking jitter variance than the latter at the expense of greater hardware complexity. Next, the performance of spread spectrum systems in jamming environments was considered. Both barrage noise jamming performance and optimized jammer performance were considered. The use of coding to combat the deleterious effects of jamming was considered. Finally, the performance of code-division multiple-access systems was briefly surveyed, with the near-far problem illustrated by example.

REFERENCES

[1] R. L. Peterson, R. E. Ziemer, and D. E. Borth, *Introduction to Spread Spectrum Communications*, Upper Saddle River, NJ: Prentice Hall, 1995.

[2] M. K. Simon, J. K. Omura, R. A. Scholtz, and B. K. Levitt, *The Spread Spectrum Handbook*, revised edition, New York: McGraw Hill, 1994.

[3] K. S. Zigangirov, *Theory of Code Division Multiple Access Communication*, New York: Wiley/IEEE Press, 2004.

[4] R. A. Dillard and G. M. Dillard, *Detectability of Spread Spectrum Signals*, Norwood, MA: Artech House, 1989.

[5] TIA/EIA Interim Standard-95, "Mobile Station—Base Station Compatibility Standard for Dual-Mode Wideband Spread Spectrum Cellular System," July 1993.

[6] J. Geier, *Wireless LANs*, 2nd edition, Indianapolis, IN: Sams Publishing, 2001.

[7] D. V. Sawate and M. B. Pursley, "Cross-Correlation Properties of Pseudorandom and Related Sequences," *Proc. IEEE*, vol. 68, pp. 593–619, May 1980.

[8] G. L. Stuber, *Principles of Mobile Communication*, 2nd ed., Boston: Kluwer Academic Publishers, 2001.

[9] P. V. Kumar, T. Helleseth, A. R. Calderbank, and A. R. Hammons, Jr., "Large Families of Quaternary Sequences with Low Correlation, *IEEE Trans. Inf. Theory*, vol. 42, pp. 579–592, Mar. 1996. doi:10.1109/18.485726

[10] A. R. Hammons, Jr. and P. V. Kumar, "On Recent 4-Phase Sequence Design for CDMA," *IEICE Trans. Commun.*, vol. E76-B, pp. 804–813, Aug. 1993.

[11] H. Urkowitz, "Energy Detection of Unknown Deterministic Signals," *Proc. IEEE*, vol. 55, pp. 523–531, April 1967.

[12] A. Polydoros and C. Weber, "A Unified Approach to Serial Search Spread- Spectrum Code Acquisition—Part II: A Matched Filter Receiver," *IEEE Trans. on Commun.*, vol. COM-32, pp. 550–560, May 1984.

[13] B.K. Levitt, "Effect of Modulation Format and Jamming Spectrum on Performance of Direct Sequence Spread Spectrum Systems," *Conf. Record, IEEE National Telecommunications Conf.*, pp. 3.4.1–3.4.5, 1980.

[14] R. Singh, "Performance of Direct Sequence Spread Spectrum Systems with Long Period and Short Period Code Sequences," *Conf. Record, IEEE International Conf. on Commun.*, pp. 45.2.1–45.2.5, 1981.

[15] S.W. Houston, "Modulation Techniques for Communication: Part1. Tone and Noise Jamming Performance of Spread Spectrum *M*-ary FSK and 2, 4-ary DPSK Waveforms," *Conf. Rec.*, NAECON, pp .51–58, 1975.

[16] R. E. Ziemer and R. L. Peterson, *Introduction to Digital Communication,* 2nd ed., Upper Saddle River, NJ: PrenticeHall, 2000.

[17] M. B. Pursley, "Performance Evaluation of Phase-Coded Spread-Spectrum Multiple-Access Communication," *IEEE Trans. on Commun.,* vol. COM-25, pp. 800–803, Aug. 1977. doi:10.1109/TCOM.1977.1093916

[18] K.B. Letaief, "Efficient Evaluation of the Error Probabilities of Spread-Spectrum Multiple-Access Communications," *IEEE Trans. on Commun.,* vol. 45, pp. 139–246, Feb.1997. doi:10.1109/26.554372

[19] R. Michael Buehrer, *Code Division Multiple Access (CDMA)*, San Rafael, CA: Morgan and Claypool, 2006.

Author Biography

Rodger E. Ziemer received the BSEE, MSEE, and Ph.D. degrees from the University of Minnesota in 1960, 1962, and 1965, respectively. After serving in the U.S. Air Force from 1965 to 1968, he joined the University of Missouri–Rolla in 1968 where he stayed until 1983, having been promoted through the ranks to Professor. He joined the Electrical and Computer Engineering (ECE) Department of the University of Colorado at Colorado Springs (UCCS) in January 1984 where he was Professor and Chairman of ECE until 1993 and then Professor from September 1993 till now. In August 1998, he went on leave to the National Science Foundation where he served as Program Director for Communications Research until August 2001, and then returned to UCCS. He has spent intermittent periods on leave or sabbatical to various universities and industrial concerns, including Motorola Government Electronics Group in 1980–81 and in 1991, Motorola Corporate Research Laboratories in the summer of 1995, Motorola Cellular Infrastructure Group Applied Research Laboratories in the summer of 1997, University of California at San Diego in February 1998, and Virginia Technical and State University in June 1998. He was also a Visiting Professor, Iasi Polytechnic Institute, Iasi, Romania, May–June 1993, and again in May–June 1996, from which he received a Doctorate Honoris Causa. He has published several papers in his areas of research interest, principally in digital communications. He has authored and co-authored several books, including *Introduction to Digital Communications* (2nd ed.), Prentice Hall, 2001 (with R. L. Peterson), *Signals and Systems: Continuous and Discrete* (4th ed.), Prentice Hall, 1998 (with W. H. Tranter and D. R. Fannin), *Principles of Communications: Systems, Modulation, and Noise* (5th ed.), John Wiley & Sons, 2002 (with W. H. Tranter), *Introduction to Spread Spectrum Communications*, Prentice Hall, 1995 (with R. L. Peterson and D. Borth), and *Introduction to Engineering Probability and Statistics*, Prentice Hall, 1997.

